Lecture Notes in Mathematics

Edited by A. Dold and B. Eckmann

687

Algebraic Geometry

Proceedings, Tromsø Symposium, Norway,
June 27 – July 8, 1977

Edited by Loren D. Olson

Springer-Verlag
Berlin Heidelberg New York 1978

Editor
Loren D. Olson
Mathematics Department
University of Tromsø
N-9001 Tromsø/Norway

AMS Subject Classifications (1970): 14 C15, 14 H99, 14 J10, 14 A05, 14 N10

ISBN 3-540-08954-3 Springer-Verlag Berlin Heidelberg New York
ISBN 0-387-08954-3 Springer-Verlag New York Heidelberg Berlin

Printed in Germany

Printing and binding: Beltz Offsetdruck, Hemsbach/Bergstr.
2141/3140-543210

PREFACE

From June 27 to July 8, 1977, a symposium on algebraic geometry was held at the University of Tromsø, Norway. The lectures delivered there were primarily focused on the two topics of intersection theory and space curves. The contributions presented here are in large measure based on talks given at the symposium.

The symposium was supported financially by the Norwegian Research Council for Science and Humanities (NAVF) and by the University of Tromsø. We wish to express our thanks to both for their support.

Loren D. Olson

TABLE OF CONTENTS

DEFINING ALGEBRAIC INTERSECTIONS

William Fulton and Robert MacPherson

(Brown University, Providence, RI)

Contents

Suppose V and W are subvarieties of dimension v and w of a nonsingular algebraic variety X of dimension n. We are going to study the following:

Intersection Problem Find an equivalence class $V \cdot W$ of algebraic $v + w - n$ cycles which represents the algebraic intersection of V and W.

The traditional approach to this problem is to solve it directly only for the particular case where V and W meet so as to determine a unique intersection cycle. Then for general cycles V and W, the problem is solved by first moving one of them in a rational family so that this particular case applies. Since the motion involved is not unique, the intersection cycle so constructed is not uniquely determined. However, it is defined up to rational equivalence in X by this procedure.

Our object here is to consider the intersection problem directly for arbitrary V and W without moving them first. The aim is to define $V \cdot W$ within as small an equivalence class as possible.

1. When is V·W uniquely defined?

In this section, as background, we review the traditional approach to the intersection problem: to give sufficient conditions that $V \cdot W$ be a uniquely defined cycle, and to construct it under those conditions. We also review the analogous situation in topology and analysis.

Geometry

V and W are said to intersect properly if each irreducible component Z_i of $V \cap W$ has dimension $v + w - n$. If V and W intersect properly, then $V \cdot W$ is uniquely determined as a linear combination $\sum m(Z_i) Z_i$ of the components Z_i. In this case the intersection problem becomes one of determining the coefficients $m(Z_i)$, called the intersection multiplicities.

Any subscheme S of X of pure dimension i determines a canonical algebraic cycle $[S]$: if Z_i are the irreducible components of S and z_i are their generic points,

$$[S] = \sum \text{length}_{\mathcal{O}_{z_i}} (\mathcal{O}_S)_{z_i} \cdot Z_i$$

If V and W intersect properly, a naive guess would be that $V \cdot W = [V \cap W]$, i.e. that

$$m(Z_i) = \text{length}_{\mathcal{O}_{z_i}} \left(\mathcal{O}_V \otimes_{\mathcal{O}_X} \mathcal{O}_W\right)_{z_i}$$

This formula is adequate for curves on a surface (or, in fact, for the case that V or W is a divisor). But, as is well known, it is too large in general. (See [11] for a discussion of its failure and the history of the development of a correct formula by Severi, Weil, Chevalley, and Samuel.) Serre gave an elegant formula for the intersection multiplicities with correction terms to the naive guess:

$$m(Z_i) = \text{length}_{\mathcal{O}_{Z_i}} \left(\mathcal{O}_V \otimes_{\mathcal{O}_X} \mathcal{O}_W\right)_{Z_i} + \sum_{j=1}^{\infty} (-1)^j \text{length}_{\mathcal{O}_{Z_i}} \left(\text{Tor}_j^{\mathcal{O}_X}(\mathcal{O}_V, \mathcal{O}_W)\right)_{Z_i}$$

In a sense, this formula explains why the naive guess fails. For another explanation, see §4.

Topology

In topology, the intersection problem is to find an intersection cycle $V \cdot W$ where V and W are, say, integral simplicial v- and w-cycles on an oriented polyhedral n-manifold X. This problem was studied when $v + w = n$ by Poincaré [10] and for arbitrary dimensions by Lefschetz [9].

The cycles V and W are said to be dimensionally transverse if every simplex in $V \cap W$ has dimension $\leq v + w - n$. If V and W are dimensionally transverse then they determine a unique intersection cycle $\sum_{\sigma} m(\sigma)\sigma$ where the sum is over all $v + w - n$ dimensional simplices σ (supplied with orientations) in $V \cap W$, and $m(\sigma)$ is called the intersection multiplicity of V and W at σ.

The intersection multiplicity $m(\sigma)$ may be determined by the following procedure which is equivalent to that of Lefschetz. Choose a local coordinate chart $\mathbb{R}^n \supset U \to X$ near σ compatible with the orientation of X. Then there is a map $\phi : (V \cap U) \times (W \cap U) \to \sigma \times \mathbb{R}^n$ that sends x, y to $p(x), y - x$ where $p(x)$ in the orthogonal projection of x on σ and $y - x$ is the vector difference in $\mathbb{R}^n$. The intersection multiplicity $m(\sigma)$ then is the local degree of ϕ near $(b,0) \in \sigma \times \mathbb{R}^n$ where b is an interior point to σ.

As in geometry, if V is not dimensionally transverse to W V is traditionally first deformed to V' which is dimensionally transverse to W. The class of the cycle $V' \cdot W$ in $H_{v+w-n}(X)$ is independent of the deformation chosen. Thus for general V and W, $V \cdot W$ is defined up to homology in X.

Analysis

In analysis, the intersection problem is given two currents ω_1 and ω_2 on a compact manifold X, to find a product that corresponds to their homology product.

If ω_1 and ω_2 are both smooth differential forms, then the product is just the wedge product. De Rham in his thesis [1] showed that this corresponds to the Lefschetz intersection product (seven years before the invention of the cohomology cup product). More generally if the singular supports of ω_1 and ω_2 are disjoint, or more generally still if their wave front sets are disjoint, then a unique product exists [6]. If the currents arise from a pair of algebraic varieties intersecting properly, then the product can be determined by geometric measure theoretic methods [7]. But no general construction of a unique product encompassing all these cases exists. Perhaps the general problem is best stated this way: There exist families H_t of smoothing operators converging to the identity as $t \to 0$ but there is no canonical choice. When is $\lim_{t\to 0} H_t\omega_1 \wedge H_t\omega_2$ independent of H_t? (Or more generally, characterize the set of possible limits.

2. How well defined is $V \cdot W$ in general?

Suppose $V \cap W$ has components of dimension greater than $v + w - n$. Then how well defined can we expect the intersection cycle $V \cdot W$ to be?

First, it is too much to expect that it should be uniquely defined. For consider the case where X is the projective plane and V and W are both the same projective line contained in X. By Bezout's theorem we know that the algebraic intersection consists of a point with multiplicity one. However, there is no distinguished point in the given configuration of V, W, and X. In fact, the orbits of the symmetry group of the given data are $V \cap W$ and $X - V \cap W$. So no better answer exists than the class consisting of any point on $V \cap W$.

(Another reason it is too much to expect a uniquely defined intersection product comes from analogy with the topological situation. According to a principle which is widely believed but not precisely formulated or proved, a globally defined commutative product on the chain level that gives the usual product for $H_*(X;\mathbb{Z})$ cannot exist. A chain level product can be globally defined but not commutative, as in the cohomology cup product, or commutative but not globally defined, as in the homology intersection product. A globally defined commutative product would presumably contradict the existence of cohomology operations. The original construction of the Steenrod squares, for example, explicitly used the non-commutativity of the chain level cup product. Note that over the rationals, where no cohomology operations exist, such a product can exist and was constructed by Sullivan [14].

In §3, we will construct a product $V \cdot W$ which satisfies the following.

<u>Assertion 1</u> $V \cdot W$ is well defined in $A_{v+w-n}(V \cap W)$. That is, the algebraic intersection $V \cdot W$ lies in the physical intersection $V \cap W$ and is well defined up to rational equivalence in $V \cap W$.

Assertion 1 agrees with the traditional idea described in §1 that the intersection cycle is uniquely defined when V and W intersect properly. This is because the homomorphism of the group of algebraic $v + w - n$ cycles on $V \cap W$ to $A_{v+w-n}(V \cap W)$ is an isomorphism if and only if V intersects W properly. So for proper intersections, a well defined class in $A_{v+w-n}(V \cap W)$ is a uniquely determined algebraic cycle. Assertion 1 also agrees with the above example where $V = W =$ a projective line in the projective plane since the set of points in $V \cap W$ is exactly the set of effective cycles in a class of $A_0(V \cap W)$.

Another reason to expect Assertion 1 is that the corresponding topological statement is true. This can be seen by a slight refinement of the usual transversality techniques. For suppose V and W are cycles in a piecewise linear manifold X. Let N be a regular neighborhood of $V \cap W$ (so N deformation retracts to $V \cap W$). Now V can be deformed slightly to V' so that all motion takes place strictly inside N and V' is dimensionally transverse to W.

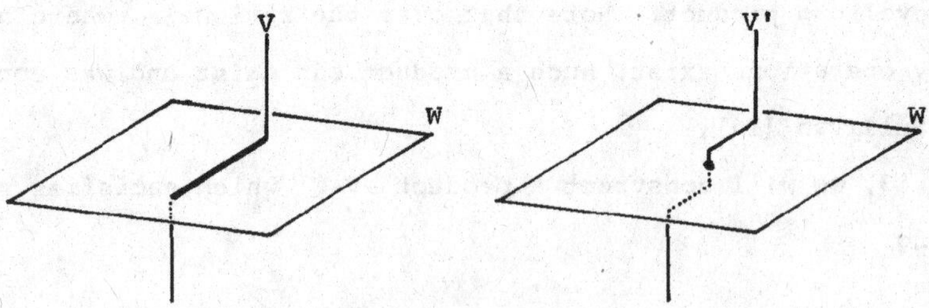

Diagram 1

Then the usual independence proof actually shows that $V' \cdot W$ is well defined in $H_{v+w-n}(N)$. But $H_{v+w-n}(N) \cong H_{v+w-n}(V \cap W)$.

Statements similar to assertion 1 were stated by Severi. He believed that a deformation technique similar to the topological one could be applied to the geometric case. If V_t is an algebraic family of algebraic cycles such that $V_{t_0} = V$ and V_t meets W properly for $t \neq t_0$, then Severi wanted to define $V \cdot W$ to be the limit of the cycles $V_t \cdot W$ as t goes to t_0. In order for this process to work, we need that

1) the limit cycle lies in $V \cap W$

2) its class in $A_*(V \cap W)$ is independent of the family V_t chosen.

Unless one adds more hypotheses, both of these assertions are false. For consider case $W = 1\{y = 0\}$, $V = 1\{x = 0\}$, $V_t = 1\{x(y - tx) + t = 0\} - 1\{y - tx = 0\}$ (properly projectivized to lie in X = projective 2 space). Note that V is the limit cycle of V_t as $t \to 0$.

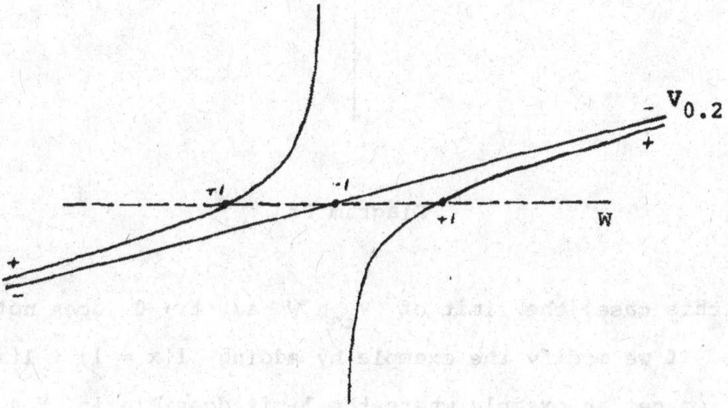

Diagram 2a

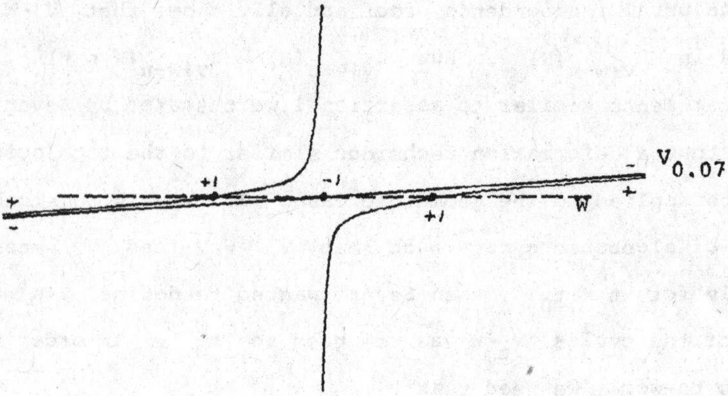

Diagram 2b

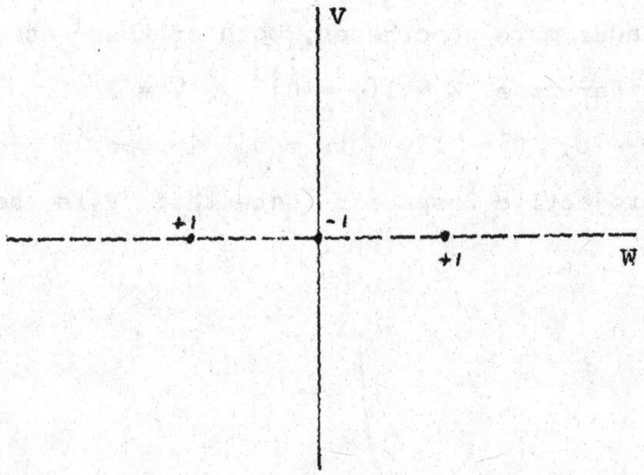

Diagram 2c

In this case, the limit of $V_t \cap W$ as $t \to 0$ does not lie in $V \cap W$. If we modify the example by adding $1\{x = 1\} + 1\{x = -1\}$ to V, we get an example where the limit does lie in $V \cap W$, but the limiting cycle is different in $A_*(V \cap W)$ from that obtained from the family $V_t = V$.

Of course this deformation is drastic: V_t is not effective (has negative coefficients) for $t \neq 0$. But there exist subvarieties with no effective deformations, such as the zero section of $\mathcal{O}(-1)$ over projective space. Algebraic geometers do not share the topologists' luxury of "small" deformations.

Bitter controversy raged over the rigor of Severi's attempts to define $V \cdot W$ in $A_{v+w-n}(V \cap W)$ by limits of families. Their intensity was probably due to the fact that the early development of algebraic intersection theory coincided with changes of standards of rigor. (Such controversies also arose over the early development of intersection theory in topology.) The important question today is not to determine who was rigorous but to resolve the often interesting geometric questions at the root of the controversies.

In this case, the question is what conditions on the family V_t produce an intersection class well defined in $A_{v+w-n}(V \cap W)$ by Severi's construction. The answer, which we state precisely in §5, is as broad as we could ask for given the above example: any family of effective algebraic cycles will do. This is striking: even if the family is not rational, the limit intersection is still determined up to rational equivalence. It seems Severi's critics expected this to be false [16].

Finally, the question arises as to whether $V \cdot W$ can be defined within a stricter equivalence than rational equivalence on $V \cdot W$. By Assertion 1, each connected component of $V \cap W$ has its own contribution to $V \cdot W$ since $A_{v+w-n}(V \cap W)$ is $\oplus A_{v+w-n}(Z_i)$ where Z_i ranges through the connected components of $V \cap W$. Severi thought that each irreducible component of $V \cap W$ should have its own contribution to $V \cdot W$, that is $V \cdot W$ should be well defined in $\oplus A_{v+w-n}(Z_i)$ where Z_i ranges through the irreducible components of

$V \cap W$ [13]. Since $\bigoplus A_{v+w-n}(Z_i)$ maps to $A_{v+w-n}(V \cap W)$, this would be a strictly stronger result.

Unfortunately, Severi's assertion is false. For let V and W be the following configurations of projective lines in the projective plane:

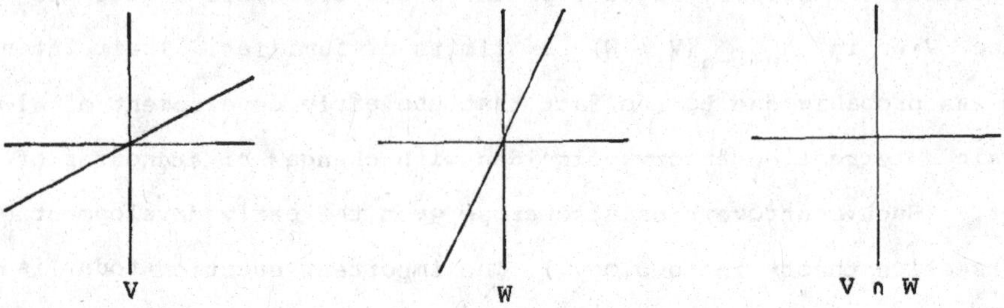

Diagram 3

$V \cap W$ has two irreducible components which are completely symmetric with respect to the given data: so they would have equal contributions to $V \cdot W$. But by Bezout's theorem, $V \cdot W$ must have total degree nine, which is odd.

The phenomenon does not depend on the reducibility of V and W as the following clearer example shows. With coordinates (x,y,z,w) on $\mathbb{P}^4$, define surfaces V and W by

$$V = \{z^3 - xy(y - 2x) = w = 0\}$$
$$W = \{w^3 - yx(x - 2y) = z = 0\}$$

Then V intersects W in the union of two lines, and the involution $(x,y,z,w) \to (y,x,w,z)$ interchanges V and W, as well as the two lines. Bezout's theorem still would require a total of nine points.

However, there is a strengthened intersection of the general type envisioned by Severi. The intersection $V \cap W$ sits naturally as a subscheme of V and of W, and therefore of $V \times W$. Let $\pi : C \to V \cap W$ be the normal cone to $V \cap W$ in $V \times W$. If C_i are the irreducible components of C, then the subvarieties $Z_i = \pi(C_i) \subset V \cap W$ may be called the distinguished varieties of the intersection.

The irreducible components of $V \cap W$ are always distinguished varieties of the intersection of V and W. But there may be more: Z_3 (the origin) is a distinguished subvariety in the example above (essentially because $V \cap W$ has nilpotent elements in its local ring at the origin).

We have the following refinement of Assertion 1.

Assertion 2 $V \cdot W$ is well defined in $\bigoplus A_{v+w-n}(Z_i)$ where the Z_i are the distinguished varieties of the intersection of V and W.

Assertion 2 is useful in applying intersection theory to certain geometric problems, where contributions to $V \cdot W$ from different distinguished varieties will have different geometric interpretations. See, for example, §7.

We do not know whether Assertion 2 can be further improved upon.

§3 The deformation

In this section we will give the construction satisfying Assertion 1 and Assertion 2 of the last section. We will omit some of the proofs, which are contained in the original papers [2] and [5].

First, the problem of intersecting V and W in X may be replaced by that of intersecting $V \times W$ and Δ (the diagonal) in $X \times X$. (This has been a key remark in most work on intersection theory to date.) We picture the situation schematically by drawing a horizontal plane representing $X \times X$ containing $V \times W$ and Δ as follows.

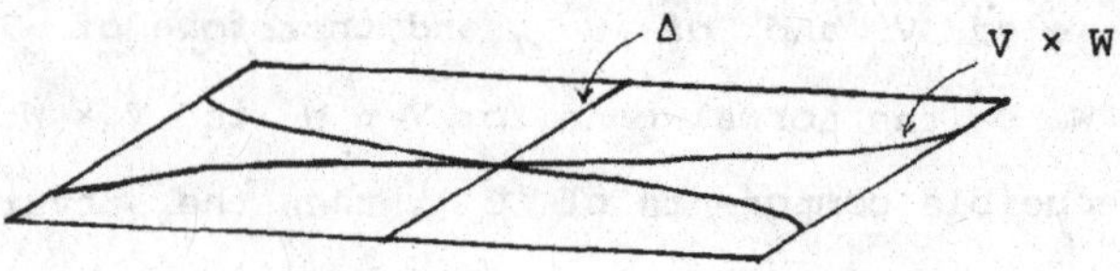

Diagram 4

Now, taking advantage of the idea that intersections are preserved under reasonable deformations, we apply the "deformation to the normal bundle" of Δ . This will deform the whole intersection problem to a more manageable one.

Let E be a vector bundle with a section s vanishing exactly on Δ as a scheme. (In general, the fiber dimension of E must be more than n for global reasons.) Let $P(E \oplus 1)$ be the projective completion of E . We may picture $P(E \oplus 1)$ as a box where the top and bottom faces, representing ∞ in $P(E \oplus 1)$, are identified with each other.

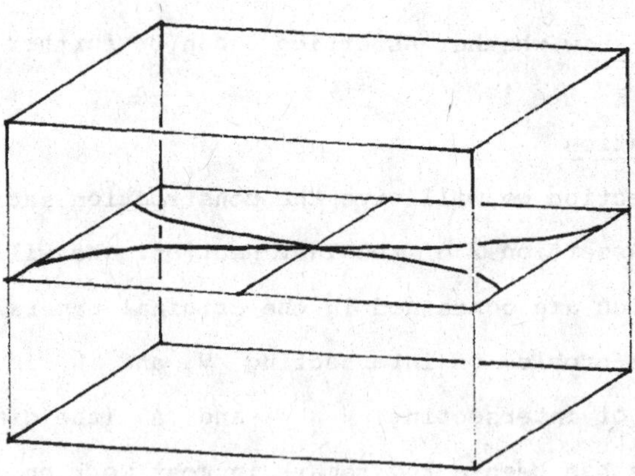

Diagram 5

Let $(X \times X)_\lambda$ and $(V \times W)_\lambda$ be the images of $V \times W$ and $X \times X$ in $E \subset P(E \oplus 1)$ by the map $\lambda \cdot s$ where λ is a field element.

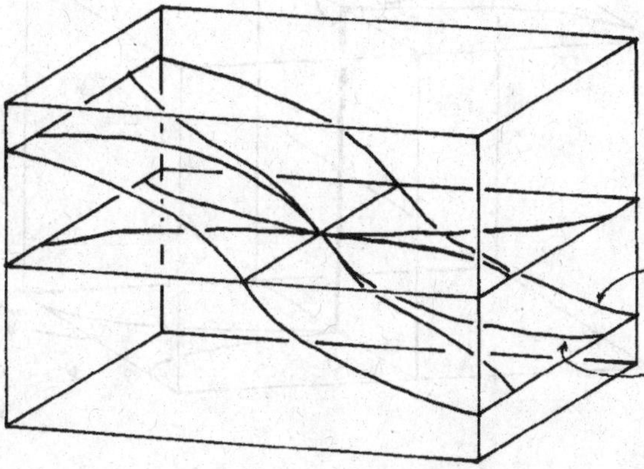

Diagram 6a

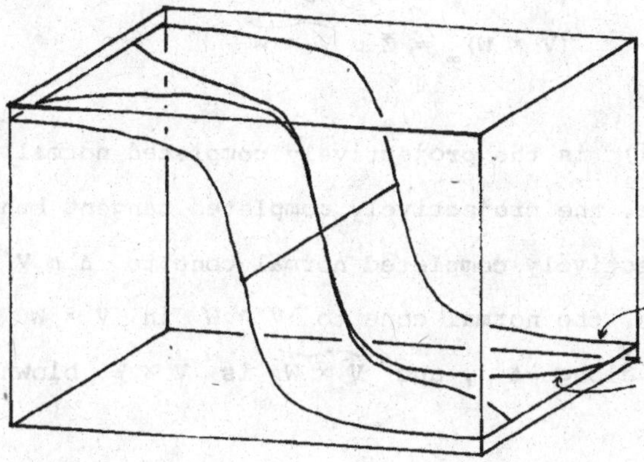

Diagram 6b

We complete these to families parametrized by the projective line and let λ go to ∞ .

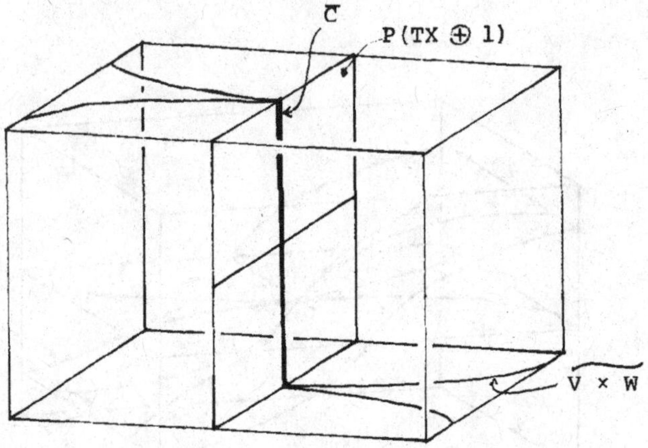

Diagram 7

Then we have the following scheme theoretic relations:

$$(X \times X)_{\infty} = P(TX \oplus 1) \cup \widetilde{X \times X}$$
$$(V \times W)_{\infty} = \bar{C} \cup \widetilde{V \times W}$$

where $P(TX \oplus 1)$ is the projectively completed normal bundle to Δ in $X \times X$ (i.e. the projectively completed tangent bundle to X), $\bar{C}$ is the projectively completed normal cone to $\Delta \cap V \times W = V \cap W$ in $V \times W$ (i.e. the normal cone to $V \cap W$ in $V \times W$), $\widetilde{X \times X}$ is $X \times X$ blown up along Δ , and $\widetilde{V \times W}$ is $V \times W$ blown up along $\Delta \cap V \times W$.

It is quite plausible that our original problem, intersecting $V \times W$ and Δ in $X \times X$, is equivalent to the deformed one, intersecting $[(V \times W)_\infty]$ and Δ in $(X \times X)_\infty$. But this is the same as intersecting $[\overline{C}]$ and Δ in $P(TX \oplus 1)$ since $\widetilde{V \times W}$ and Δ do not intersect.

This is an improvement over our original situation because $\overline{C}$ lies in $\pi^{-1}(V \cap W)$, where π is the projection of $P(TX \oplus 1)$ to X . (In a sense $\pi^{-1}(V \cap W)$ forms an algebraic substitute for the regular neighborhood N of the proof of the topological Assertion 1 in §1; π substitutes for the retraction.)

Now we can solve our problem. Move Δ to Δ' in $P(TX \oplus 1)$ so as to intersect C properly. Then calculate $[\overline{C}]\cdot\Delta'$; the element of $A_{v+w-n}(\overline{C})$ determined by $[\overline{C}]\cdot\Delta'$ is independent of the choice of Δ' . Then

$$V\cdot W = \pi_*([\overline{C}]\cdot\Delta')$$

is well defined in $A_{v+w-n}(V \cap W)$: it answers Assertion 1.

Similarly if $\overline{C}_i$ are the irreducible components of $\overline{C}$, $\overline{C}_i\cdot\Delta'$ is well defined in $A_{v+w-n}(\overline{C}_i)$ so $\pi_*(\overline{C}_i\cdot\Delta')$ is well defined in $A_{v+w-n}(Z_i)$ where $Z_i = \pi\overline{C}_i$. If $m(\overline{C}_i)$ is the multiplicity of $\overline{C}_i$ in $[\overline{C}_i]$ (see §1) then $[\overline{C}] = \sum m(\overline{C}_i)\overline{C}_i$ so

$$V\cdot W = \sum m(\overline{C}_i)\pi_*(\overline{C}_i\cdot\Delta')$$

answers Assertion 2.

§4 Application to proper intersections

The construction of $V\cdot W$ in the previous section holds, of course, whether or not V and W intersect properly. In the proper intersection case, it gives a simpler construction than Serre's of the intersection multiplicities, closely related to that of

Samuel [12].

Let V and W have proper intersection and let Z_i be the irreducible components of $V \cap W$. Then it is easily seen that $\pi^{-1}Z_i$ are the irreducible components of $\bar{C}$, i.e. they are the $\bar{C}_i$. Since $\pi_*(\pi^{-1}Z_i \cdot \Delta') = Z_i$, the multiplicity $m(Z_i)$ in $V \cdot W$ is just the multiplicity of $\pi^{-1}Z_i$ in $[\bar{C}]$, i.e. $\mathrm{length}_{\mathcal{O}_{p_i}}(\mathcal{O}_{\bar{C}})_{p_i}$ where p_i is the generic point in $\pi^{-1}Z_i$. We have the containments of schemes

$$\pi^{-1}(V \cap W) \supset \bar{C} \supset \pi^{-1}Z_i$$

which gives the inequalities on the lengths

$$\mathrm{length}_{\mathcal{O}_{p_i}}\left(\mathcal{O}_{\pi^{-1}V\cap W}\right)_{p_i} \geq \mathrm{length}_{\mathcal{O}_{p_i}}(\mathcal{O}_{\bar{C}})_{p_i} \geq \mathrm{length}_{\mathcal{O}_{p_i}}\left(\mathcal{O}_{\pi^{-1}Z_i}\right)_{p_i}$$

$$\mathrm{length}_{\mathcal{O}_{z_i}}\left(\mathcal{O}_V \otimes_{\mathcal{O}_X} \mathcal{O}_W\right)_{z_i} \geq m(Z_i) \geq 1$$

where z_i is the generic point of Z_i. In other words, this shows directly why the intersection multiplicities are always positive and also why the naive guess $V \cdot W = [V \cap W]$ gives multiplicities that are too big. It would be interesting to see whether there is a direct connection between the scheme structure of $\bar{C}$ and the higher Tor terms of Serre's formula.

One can also see directly why this construction of the intersection multiplicities agrees for complex varieties with the topological one sketched in §1. The key is to observe that in a local analytic coordinate chart $X \hookleftarrow U \subset \mathbb{C}^n$, the section of the trivial $\mathbb{C}^n$ bundle over $U \times U$ that sends (x_1, x_2) to $x_1 - x_2$ vanishes exactly on the diagonal, and so can be used in the construction of §3.

§5 Cone bundles and Segre classes

One advantage of the formula for $V \cdot W$ in §3 is that Δ' is a very special class: it is the nth Chern class of the quotient hyperplane bundle over $P(TX \oplus 1)$. This allows us to use the powerful manipulative machinery of characteristic classes to write formulas for $V \cdot W$ in terms of geometric data.

First, we generalize the problem. In the deformation procedure of §3, the diagram

$$\begin{array}{ccc} V \cap W & \hookrightarrow & V \times W \\ \downarrow & & \downarrow \\ \Delta & \hookrightarrow & X \times X \end{array}$$

may be replaced by any fiber square

$$\begin{array}{ccc} Y' & \xrightarrow{j} & X' \\ g \downarrow & & \downarrow f \\ Y & \underset{i}{\hookrightarrow} & X \end{array}$$

where X and Y are nonsingular. Then if $\pi : P(N \oplus 1) \to Y$ is the projection to Y of the completed normal bundle of Y in X, and if $\bar{C}$ is the completed normal cone of Y' in X', the deformation gives

$$X' \cdot Y = \pi_*([\bar{C}] \cdot Z)$$

where Z is equivalent in $A_*(P(N \oplus 1))$ to the image of the zero section of N.

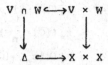

Denote by n the codimension of Y in X by m the dimension of X' , and by k and π' the map

$$\begin{array}{ccc} \bar{C} & \overset{k}{\hookrightarrow} & P(N \oplus 1) \\ & {\scriptstyle \pi'} \searrow \quad \swarrow {\scriptstyle \pi} & \\ & Y & \end{array}$$

The class of Z is $c_n(Q)$ where $Q = \pi^*(N \oplus 1)/\mathcal{O}(-1)$, so we have

$$\begin{aligned} X' \cdot Y &= \pi_*([\bar{C}] \cdot Z) \\ &= \pi_*([\bar{C}] \cdot c_n Q) \\ &= \pi'_*(k^* c_n Q \cdot [\bar{C}]) \\ &= \pi'_*(c_n(\pi'^* N \oplus 1/k^*\mathcal{O}(-1)) \cdot [\bar{C}]) \\ &= \left[\pi'_*(\pi'^* c_* N) \cdot (c_* k^* \mathcal{O}(-1))^{-1} \cdot [\bar{C}]\right]_{m-n} \\ &= \left[c_* N \cdot \pi'_*(c_* k^* \mathcal{O}(-1))^{-1} \cdot [\bar{C}]\right]_{m-n} \end{aligned}$$

where the subscript $m - n$ means take the component in dimension $m - n$.

The expression $\pi'_*(c_* k^* \mathcal{O}(-1))^{-1} \cdot [\bar{C}]$ as a class in $A_*(Y')$ depends only on the structure of C as a "cone bundle" over Y' (not on the embedding k .) It is called the Segre class of C and notated $S(C)$. Our final formula then is

$$X' \cdot Y = g_*[g^* c_* N \cdot S(C)]_{m-n}$$

(This formula makes sense for a general fiber square when i is a local complete intersection morphism if the Chern classes involved are defined as in [3]. In this case, we define

$$X' \cdot_f Y = [g^* c_* N \cdot S(C)]_{m-n}$$

This can be interpreted as a cap product of a relative cohomology class determined by Y with the homology class of X .)

In order for these formulas to be useful, one needs a theory of Segre classes of "cone bundles". (Note that our "cone bundles" need not be locally trivial. Their Segre classes lie in a homology-like theory, the Chow group, not a cohomology like ring of operators.) We sketch some rudiments of such a theory here.

<u>Proposition 1</u> If E is a vector bundle and C is a cone bundle over Y , then

$$S(E \times_Y C) = (c_* E)^{-1} \circ S(C)$$

It follows that for any vector bundle, the Segre class is the inverse of the Chern class.

A birational map of "cone bundles" is a diagram

$$\begin{array}{ccc} C' & \xrightarrow{g} & C \\ \downarrow & & \downarrow \\ Y' & \xrightarrow[f]{} & Y \end{array}$$

such that g is birational and sends generator lines onto generator lines; such a map induces a birational map on the projectivizations.

<u>Proposition 2</u> If f and g form a birational map of "cone bundles" as above, then

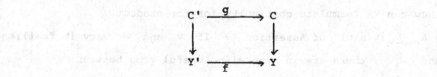

$$f_* s(C') = s(C)$$

This can be used to calculate any Segre class since any "cone bundle" blown up along its zero section is a line bundle. Also it means that in a geometric situation, the normal cone needs to be determined only up to birational equivalence.

§6 Continuity

Perhaps the most important property of algebraic intersections is continuity, which says roughly that when V and W are deformed continuously, the product $V \cdot W$ varies continuously also. If $v + w = n$, this implies in particular that the total degree of the intersection (the sum of the multiplicities of the intersection points) is constant when V and W vary in algebraic families. This form was called continuity of number by Poncelet in 1822 and was the main tool for the spectacular calculations of enumerative geometry of the 19th century. From the modern point of view, continuity is what guarantees that the intersection product defines a ring structure on $A_*(X)$.

How can we formulate continuity for the product $V \cdot W \in A_{v+w-n}(V \cap W)$ of Assertion 1? If V and W vary in families V_t and W_t, there are in general no useful maps between $A_*\left(V_{t_0} \cap W_{t_0}\right)$ and $A_*\left(V_{t_1} \cap W_{t_1}\right)$. However, there will be a group which specializes to both of them.

Let $\tilde{V}$ and $\tilde{W}$ be subvarieties of dimension $v + 1$ and $w + 1$ of $X \times D$ where D is a nonsingular algebraic curve. For $t_0 \in D$ form the fiber square

$$\begin{array}{ccc} V_{t_0} \cap W_{t_0} & \hookrightarrow & \tilde{V} \cap \tilde{W} \\ \downarrow & & \downarrow f \\ t_0 & \longrightarrow & D \end{array}$$

Then there is a specialization map

$$A_*(\tilde{V} \cap \tilde{W}) \longrightarrow A_*\left(V_{t_0} \cap W_{t_0}\right)$$

which sends α to $\alpha \cdot_f t_0$ (see §5). The following proposition follows from the theory of such specializations developed in [4] [2] [17].

<u>Proposition</u>

$$V_{t_0} \cdot W_{t_0} = (\tilde{V} \cdot \tilde{W}) \cdot_f t_0$$

This proposition answers the geometric question raised by Severi's work on defining $V \cdot W$ by a limiting procedure. (§2)

The problem of continuity for more refined $V \cdot W$ such as that of Assertion 2 appears to be much more complicated.

§7 <u>A classical enumeration problem</u>

An early motivation for developing intersection theory was to solve problems in enumerative geometry. The ambient space X is the parameter space for a family of curves (or surfaces), while the subvarieties to be intersected parametrize members of the family which satisfy special conditions. One of the simplest such problems, where the conditions involve more than points or lines, is the following:

<u>How many plane conics are tangent to five given plane conics?</u>

This question was raised by Steiner in 1848. He and Bischoff (1859) at first gave the incorrect answer of 7776. The correct answer of 3264, when the five given conics are in general position, was found by de Jonguières and Chasles (1864). Kleiman has reported on the history of this and related problems, and presented a modern justification of Chasles's method [8].

The refined intersection theory we have been discussing can be used to illuminate such questions. We illustrate this by working out the above example.

A plane conic is given in homogeneous coordinates by a second degree equation

$$ax^2 + bxy + cy^2 + dxz + eyz + fz^2 = 0 \quad .$$

The six coefficients are determined up to multiplication by a scalar, so the set of all conics is identified with $\mathbb{P}^5$. This includes the conics consisting of two lines, or one double line, as well as the non-singular conics.

Two conics are tangent if they intersect in fewer than four distinct points (or if they coincide). If one non-singular conic D is fixed, the set of conics tangent to D form a hypersurface H_D in $\mathbb{P}^5$ of degree 6. This may be visualized by choosing an ellipse as shown for D, and choosing a line in $\mathbb{P}^5$--i.e. a pencil of conics--to be the set of circles tangent to the y-axis at the point P. The six circles illustrated represent the intersections of this line with H_D.

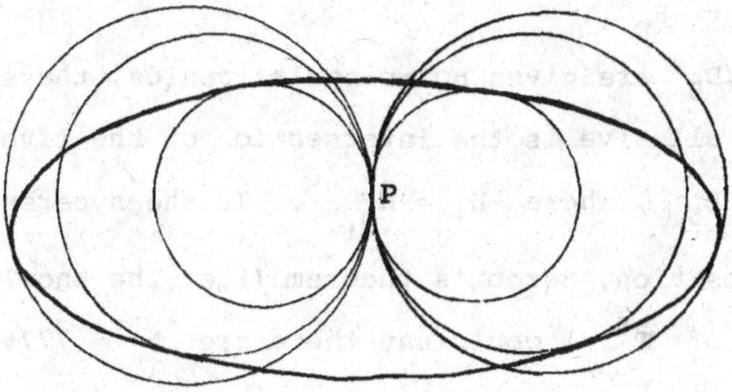

Here is one way to write down the equation for H_D . Since all non-singular conics are projectively equivalent, we may take D to be the conic $y^2 = xz$, and choose an isomorphism $f : \mathbb{P}^1 \to D$ by $f(s,t) = (s^2, st, t^2)$. The condition for a conic $ax^2 + bxy + cy^2 + dxz + eyz + fz^2 = 0$ to be tangent to D is that $as^4 + bs^3t + cs^2t^2 + ds^2t^2 + est^3 + ft^4$ not have four distinct roots. The discriminant of this polynomial,

$$\det\begin{pmatrix} 1 & b & c+d & e & f & 0 & 0 \\ 0 & a & b & c+d & e & f & 0 \\ 0 & 0 & a & b & c+d & e & f \\ 4 & 3b & 2(c+d) & e & 0 & 0 & 0 \\ 0 & 4a & 3b & 2(c+d) & e & 0 & 0 \\ 0 & 0 & 4a & 3b & 2(c+d) & e & 0 \\ 0 & 0 & 0 & 4a & 3b & 2(c+d) & e \end{pmatrix}$$

is the equation for H_D .

If $D_1, \ldots, D_5$ are given non-singular conics, the set of conics tangent to all five is the intersection of the five hypersurfaces $H_1 \cap \ldots \cap H_5$, where $H_i = H_{D_i}$. If the hypersurfaces were in general position, Bezout's theorem (i.e. the knowledge of the intersection ring of $\mathbb{P}^5$) would say there are $6^5 = 7776$ conics in this intersection.

These hypersurfaces are never in general position, however. They all contain the Veronese surface V of all double lines; V is isomorphic to the projective plane by the correspondence

$$(a,b,c) \rightarrow (ax + by + cz)^2 .$$

(If four of the D_i were tangent to the same line, there would be another curve in $\cap H_i$, as well as isolated points representing the desired conics.)

Consider the fibre square

$$\begin{array}{ccc} H_1 \cap \ldots \cap H_5 & \hookrightarrow & \mathbb{P}^5 \\ \downarrow g & & \downarrow \delta \\ H_1 \times \ldots \times H_5 & \overset{i}{\hookrightarrow} & \mathbb{P}^5 \times \ldots \times \mathbb{P}^5 \end{array}$$

where δ is the diagonal imbedding. The normal bundle N to the imbedding i restricts to the direct sum of five copies of $\mathcal{O}(6)$ on $H_1 \cap \ldots \cap H_5$.

The intersection theorem discussed in 5 says that the class

$$[g^*c(N) \cdot S(\cap H_i, \mathbb{P}^5)]_0$$

in $A_0(\cap H_i)$ maps to the intersection product $H_1 \cap \ldots \cap H_5 = 6^5$ in $A_0\mathbb{P}^5 = \mathbb{Z}$. Here $S(\cap H_i, \mathbb{P}^5)$ is the Segre class of the normal cone to $\cap H_i$ in $\mathbb{P}^5$. We will compute the contribution to this intersection product from each component of $\cap H_i$, under mild assumptions on the five given conics.

<u>Proposition 1. If no three of the five given conics pass through a point, then the Veronese V is a connected component of $\cap H_i$, and the contribution to the intersection product in $A_0V = \mathbb{Z}$ is</u> 4512.

<u>Proof</u>. Let $\pi : \tilde{\mathbb{P}}^5 \to \mathbb{P}^5$ be the blow-up of $\mathbb{P}^5$ along V, $E = \pi^{-1}(V)$ the exceptional divisor. If a point in V is represented by a double line L^2, then the fibre of E over this point may be identified with the set of positive divisors of degree two on L. If H_D is the hypersurface of conics tangent to a non-singular conic D, then

$$\pi^* H_D = 2E + G_D$$

where G_D is a divisor on $\tilde{\mathbb{P}}^5$. A point in E represented by a double line L and a divisor of degree two on L is in G_D if and only if at least one of the points of the divisor is on D:

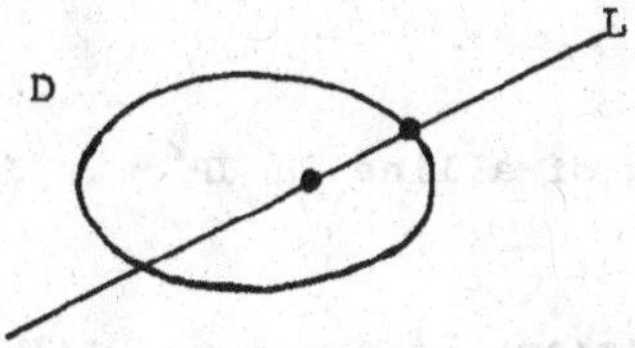

Diagram 7

This is not the place to develop a general method to explain this; we leave the reader to verify these facts by a direct calculation using the equation for H_D.

Now let H_i be the hypersurface of conics tangent to D_i, $i=1, \ldots ,5$, and $\pi^*H_i = 2E + G_i$. By assumption, two points cannot meet all five conics, so $\cap G_i \cap E$ is empty. It follows that the divisor $2E$ is, scheme-theoretically, a connected component of $\pi^{-1}(\cap H_i) = \cap \pi^*H_i$.

From the birational invariance of the Segre class (§5), the contribution of $S(\cap H_i, \mathbb{P}^5)$ in the connected component V of $\cap H_i$ is therefore

$$\eta_*(c(\mathcal{O}(2E))^{-1} \cdot 2E) \quad ,$$

where $\eta : E \to V$ is the projection induced by π. (In this and the following formulas, if f is a morphism from a scheme A to a scheme B, and F is a vector bundle on B, $c_*(F) \cdot A$ denotes the result of capping the total Chern class of f^*F with the fundamental class of A.) Now

$$\begin{aligned} S(V,\mathbb{P}^5) &= \eta_*(c_*(\mathcal{O}(E))^{-1} \cdot E) \\ &= c_*\left(T_{\mathbb{P}^5}/T_V\right)^{-1} \cdot V \\ &= (1 + h)^3/(1 + 2h)^6 \\ &= 1 - 9h + 51h^2 \qquad \text{in } A_*V = A_*\mathbb{P}^2 \end{aligned}$$

where h is the class of a line in $\mathbb{P}^2$. It follows that

$$\eta_*(c_*(\mathcal{O}(2E))^{-1} \cdot 2E) = 8 - 144h + 1632h^2$$

so the contribution to the intersection product in A_0V is

$$\left[(1 + 6(2h))^5(8 - 144h + 1632h^2)\right]_0 = 4512h^2 \quad ,$$

as desired.

Proposition 2. If no two of the five given conics are tangent to each other, and no three of them are tangent to any line, then the intersections of $H_1 \cap \ldots \cap H_5$ outside the Veronese are all isolated points corresponding to non-singular conics. If C is a non-singular conic, the intersection number of $H_1 \cap \ldots \cap H_5$ at the point corresponding to C is

$$\prod_{i=1}^{5} (4 - \mathrm{card}(C \cap D_i)) \quad .$$

Proof: The projective tangent space to the point $\{C\}$ corresponding to C in $\mathbb{P}^5$ may be identified with the set of positive divisors of degree four on C ; the divisors containing a point P on C form a hyperplane L_P . For another non-singular conic D the projective tangent cone to H_D at $\{C\}$ is

$$\prod_{P \in C \cap D} L_P^{r_P}$$

where $r_P = I(P, C \cap D) - 1$, i.e. r_P is one less than the intersection number of curves C and D at P ; this may be calculated using the equation for H_D . If no two D_i are tangent, the intersection of the projective tangent cones of $H_1, \ldots, H_5$ is empty. For such hypersurfaces, the intersection number is the product of the multiplicities (a general fact which may be proved as in the previous proposition), which gives the answer

$$\prod_{i=1} \left(\sum_{P \in C \cap D_i} (I(P, C \cap D_i) - 1) \right)$$

which is the desired result.

If five given conics satisfy the hypotheses of the two propositions, there are therefore $7776 - 4512 = 3264$ non-singular conics which are tangent to all of them, provided each is counted with the prescribed multiplicity.

With this explicit description of multiplicities, a simple dimension count shows that these multiplicities are all one when the five given conics are in general position, so there are 3264 distinct non-singular tangent conics in the general case. For if we say that C and D are <u>multiply tangent</u> if

$$\sum_{P \in C \cap D} (I(P, C \cap D) - 1) > 1 \quad ,$$

the set of conics D which are multiply tangent to a fixed non-singular conic C form a three-dimensional algebraic set J_C in $\mathbb{P}^5$. If $\mathbb{P}^5_0$ is the open subvariety of non-singular conics, let S_k be the subset of $\mathbb{P}^5_0 \times \mathbb{P}^5 \times \ldots \times \mathbb{P}^5$ consisting of those $(C, D_1, \ldots, D_5)$ such that C is tangent to each D_i, and C is multiply tangent to D_k. Then S_k is a closed algebraic subset, and the fibre of the projection from S_k to $\mathbb{P}^5_0$ over the point C is a product of 4 H_C's and one J_C. Therefore S_k is 24-dimensional, so the projections of $S_1 \cup \ldots \cup S_5$ to $\mathbb{P}^5 \times \ldots \times \mathbb{P}^5$ must be contained in a proper algebraic subset, as desired. (See [8] for a proof in this generic case without calculating the intersection multiplicities.)

Similar methods can be used to find the contributions when the five conics are in the special positions not considered in propositions 1 and 2. (The above arguments are valid in all characteristics but $p = 2$. In characteristic 2, discriminants are squares, so the hypersurfaces H_i of conics tangent to a non-singular conic D_i are of degree 3, $\pi^*H_i = 2E + G_i$ as before, and a similar calculation shows that 192 of the $3^5 = 243$ intersections of Bezout's theorem are accounted for by the Veronese, leaving 51 non-singular conics tangent to the five given conics, provided they are counted correctly; this result was obtained by Vainsencher in the generic case ([15])).

We believe this refined intersection product will be useful both conceptually and computationally in working out some of Schubert's more challenging numerative assertions. We hope that modern solutions to these problems will include precise multiplicities as in the above propositions, as well as providing the generic solution.

<u>Correction</u>. The conclusion of Proposition 1 is valid only if there is no line with two points on it such that each of the five conics either contains one of the points or is tangent to the line; such a configuration belongs to the G_i constructed in the proof. In Proposition 2 one should also rule out the case where there are two lines each tangent to two of the conics and intersecting on the fifth conic, unless one wishes to count such singular solutions.

Bibliography

1. G. de Rham, Sur l'analysis situs des variétés à n dimensions, These faculté des Science de Paris, Gauthier-Villars, Paris, (1931).

2. W. Fulton and R. MacPherson, Intersecting cycles on an algebraic variety, Real and Complex Singularities Oslo, 1976, Sijthoff and Noordhoff, 179-197.

3. W. Fulton, Rational equivalence for singular varieties, Publ. Math. I.H.E.S. no. 45 (1975), 147-165.

4. W. Fulton, to appear.

5. H. Gillet, thesis, Harvard University, 1978.

6. V. Guillemin and S. Sternberg, Geometric Asymptotics, Math. Surveys no. 14, Amer. Math. Soc. 1977, p. 328.

7. R. M. Hardt, Slicing and intersection theory for chains associated with real analytic varieties, Acta Mathematica 129(1972) 75-136.

8. S. Kleiman, Chasles's enumerative theory of conics. A historical introduction. Aarhus University Preprint Series 1975/76 No. 32, Aarhus Denmark, to appear in an M.A.A. volume on algebraic geometry.

9. S. Lefschetz, Intersections and transformations of complexes and manifolds, Trans. A.M.S. 28(1926), 1-49.

10. H. Poincaré, Oeuvres, Tome VI, Gauthier-Villars, p. 218.

11. P. Samuel, Sur l'historie du quinzieme probleme de Hilbert, Gazette des Mathematiciens, Oct. 1974, p. 22-32.

12. P. Samuel, La notion de multiplicité en algèbre et en géométrie algébrique, J. Math. pures appl., 30, 1951, p. 159-274.

13. F. Severi, Über die Grundlagen der Algebraischen Geometrie, Abh. math. Sem. Hamburg Univ. vol 9, 1933, p. 335-364.

14. D. Sullivan, Infinitesimal computations in topology, Publ. Math. I.H.E.S. 47, 1977.

15. I. Vainsencher, Conics in characteristic 2, preprint, to appear in Compositio Math.

16. B. L. Van der Waerden, The theory of equivalence systems of cycles on a variety, Symposia Mathematica V, Istituto Nazionale di Alta Mathematica (1971), 255-262.

17. J.-L. Verdier, Le théorème de Riemann-Roch pour les intersections complètes, Astérisque 36-37 (1976), 189-228.

GENRE DES COURBES DE L'ESPACE PROJECTIF

par

Laurent GRUSON (Lille 1) et Christian PESKINE (Oslo)

Introduction.

Pour quels entiers positifs d et g existe-t-il une courbe lisse connexe de degré d et de genre g dans $\mathbb{P}^3$ (1) ? Halphen annonce dans son mémoire sur la classification des courbes gauches une solution complète de ce problème. Si plusieurs lois générales apparaissent clairement dans cette étude, d'autres sont mêlées à des considérations qui en rendent l'interprétation hasardeuse. Le problème abordé ici est essentiellement la détermination du genre maximum des courbes lisses connexes de degré d , non contenues dans une surface de degré < s , pour s donné. Le premier résultat dans cette direction est le suivant, impliqué par la classification d'Halphen:

Soit C une courbe lisse connexe, de degré d , dans $\mathbb{P}^3$ et soit s un entier positif tel que $s(s-1) < d$. Alors si C n'est pas contenue dans une surface de degré < s , le genre g de C vérifie l'inégalité:

$$g \leq 1 + d/2(s + d/s - 4) - r(s-r)(s-1)/2s ,$$

avec $0 \leq r < s$ et $d + r \equiv 0$ (modulo s), l'égalité ayant lieu si et seulement si C est liée à une courbe plane de degré r , par des surfaces de degrés respectifs s et $(d+r)/s$.

La démonstration d'Halphen repose sur la construction décrite au chapitre 3 de son mémoire. Mais d'une part elle s'appuie sur des

(1) Espace projectif de dimension 3 sur un corps de caractéristique nulle algébriquement clos.

arguments de position générale non démontrés et d'autre part pour $s \geq$ Halphen considère explicitement comme évident ([6], p.402) que pour d fixé, la borne supérieure des genres des courbes, situées sur une surface de degré s et non sur une surface de degré $s-1$, est une fonction décroissante de s (cette assertion nous semble au contraire la principale difficulté du théorème). Pour $s=2$, la démonstration d'Halphen est complète; c'est alors dans $\mathbb{P}^3$ la majoration du genre des courbes gauches fournie plus tard par Castelnuovo dans $\mathbb{P}^r$.

Les hypothèses de position générale de cette très belle démonstration nous paraissant inabordables, nous avons voulu en isoler le contenu algébrique. Il reste le résultat suivant:

Théorème de spécialité: Soit C une courbe intègre de degré d dans $\mathbb{P}^3$, non contenue dans une surface de degré $< s$. Pour tout entier $n \geq s+d/s-4$, la série $O_C(n)$ est spéciale si et seulement si $n = s+d/s-4$ et C est intersection complète de surfaces de degrés respectifs s et d/s.

Lorsque C est section d'un fibré de rang 2, on retrouve un résultat de Barth ([1], Cor.1 du th.3) et la majoration du genre énoncée plus haut.

La démonstration de ce théorème fait l'objet du paragraphe 1 de cette rédaction.

Au paragraphe 2, on complète la classification des courbes arithmétiquement normales de $\mathbb{P}^3$ fournie par Ellingsrud ([4]), en déterminant une condition nécessaire et suffisante pour qu'une fonction numérique soit la fonction de Hilbert d'une courbe arithmétiquement normale.

Au paragraphe 3, après avoir remarqué qu'une section plane générale d'une courbe intègre de $\mathbb{P}^3$ a la même postulation que la section plane d'une courbe arithmétiquement normale de même degré, on démontre la majoration du genre énoncée plus haut, au moyen d'un théorème de position générale dû à Laudal ([8]). Comme chez Castelnuovo ([3]) et chez Harris ([7]), la démonstration passe par une évaluation du maximum en tout $n \geq 0$ de la fonction $h^1(O_C(n)) - h^1(O_C(n+1))$.

Au paragraphe 4, quelques examples de classification sont traités.

§ 1. Le théorème de spécialité.

Soit V une variété intègre de codimension deux de $\mathbb{P}^r = \mathrm{Proj}(k[X_o,\ldots,X_r])$ $(r \geq 3)$. On suppose que l'origine 0 , de coordonnées $(0,0,\ldots,0,1)$, n'est pas un point de V et que la projection de sommet 0 dans l'hyperplan à l'infini $\mathbb{P}^{r-1}$ induit un morphisme birationnel de V sur son image $\overline{V}$. L'idéal de $\overline{V}$ dans $O_{\mathbb{P}^{r-1}}$ est alors défini par un polynôme $F \in k[X_o,\ldots,X_{r-1}]$, homogène de degré $d = \deg(V)$. Nous allons décrire sommairement, dans cette situation, une construction effectuée par Halphen lorsque $r=3$ ([6],chap.3).

Suivant Halphen, nous noterons (h_o) la sous-variété de $\overline{V}$ dont l'idéal est le conducteur de O_V dans $O_{\overline{V}}$. Soit I le plus grand idéal gradué de $k[X_o,\ldots,X_{r-1}]$ définissant (h_o) , et soit u_o un élément homogène, non nul, de degré minimum de I . Comme le conducteur de O_V dans $O_{\overline{V}}$ est isomorphe à $\omega_V(r-d)$, où ω_V est un faisceau dualisant sur V , le degré de u_o est, en vertu du théorème de dualité, $d-r-e(V)$, où $e(V) = \max\{i \text{ tels que } H^{r-2}(O_V(i)) \neq 0\}$. D'autre part I/F est par définition muni d'une structure de module sur $k[X_o,\ldots,X_r]$; on peut donc choisir, pour tout entier $i \geq 1$, un élément u_i de I , homogène de degré $d+i-r-e(V)$, dont la classe modulo F est le produit de la classe de u_o par X_r^i . On forme alors la matrice persymétrique $(u_{i+j})_{i,j\geq 0}$. Cette matrice est de rang un modulo F , donc pour tout entier $p \geq 1$, ses p-mineurs sont divisibles par F^{p-1} , d'après le lemme suivant:

Lemme 1.1: Soient A un anneau, P un idéal premier de A et $(a_{ij})_{1\leq i,j\leq m}$ une matrice, à coefficients dans A , de

<u>rang $\leq m'$ modulo P ; on a $\text{dét}((a_{ij})) \in P^{(m-m')}$ (puissance symbolique).</u>

On peut supposer A local d'idéal maximal P . On écrit alors $(a_{ij}) = (b_{ij}) + (c_{ij})$, où $(b_{ij})_{i,j}$ est de rang m' et $c_{ij} \in P$ pour tout (i,j). Le lemme se déduit de l'égalité $\text{dét}((a_{ij})) = \pm \sum_{I,J} B_{I,J} \cdot C_{CI,CJ}$ où $B_{K,L}$, resp. $C_{K,L}$, est le mineur de (b_{ij}) , resp. (c_{ij}) , défini par les lignes d'indice $\in K$ et les colonnes d'indice $\in L$. Nous pouvons maintenant démontrer une inégalité reliant

$e = e(V) = \max\{n$ tels que $H^{r-2}(O_V(n)) \neq 0\}$, et

$s = s(V) =$ minimum des degrés des hypersurfaces contenant V .

<u>Théorème 1.2 (de spécialité):</u> <u>Si $P(X)$ est le polynôme $X^2 - (e+r+1)X + d$, on a $s \leq [P(n)]_+ + n$ pour tout entier positif. De plus lorsque $P(n) \leq 0$, l'égalité a lieu si et seulement si V est intersection complète d'hypersurfaces de degrés respectifs s et d/s</u> .

Pour tout couple d'entiers positifs (p,q) , considérons la matrice

$$M_{p,q} = \begin{pmatrix} u_o, & u_1, & \cdots & , u_{q-1} \\ u_1 & & & \\ \vdots & & & \\ u_{p-1}, & u_p, & \cdots & , u_{p+q-2} \end{pmatrix}$$

Nous aurons besoin du résultat suivant dont une partie est par exemple démontrée dans ([5], chap 15, § 11, Folg. 2).

<u>Lemme 1.3:</u> <u>Soient m et n deux entiers positifs tels que $m \leq n$, que rang $(M_{m,n}) = m$ et que $\text{dét}(M_{m+1,m+1}) = \ldots = \text{dét}(M_{n,n}) = 0$. Alors $M_{m+1,n}$ est de rang m , et si $(\delta_o, \ldots, \delta_m)$ est l'unique relation, à coeffici-</u>

ents homogènes de p.g.c.d. un dans $k[X_o,\dots,X_{r-1}]$, entre les lignes de $M_{m+1,n}$, il existe un polynôme non nul B tel que $\text{dét}(M_{m,m}) = B \cdot F^{m-1} \cdot \delta_m^{n-m+1}$.

Faisons une récurrence sur $m-n$. Pour $m-n=0$, il est évident que $M_{m+1,n}$ est de rang m et la deuxième partie de l'énoncé se déduit du fait que les mineurs maximaux de $(M_{m+1,m})$ forment une relation entre les lignes de $M_{m+1,m}$. Supposons maintenant $n>m$. Si $M_{m+1,n}$ est de rang $m+1$, alors par hypothèse de récurrence $M_{m+2,n}$ est de rang $m+1$, et si $(\omega_o,\dots,\omega_{m+1})$ est l'unique relation à coefficients homogènes, de p.g.c.d. un, dans $k[X_o,\dots,X_{r-1}]$ entre les lignes de $M_{m+2,n}$, ou a $\omega_{m+1}=0$ car $\text{dét}(M_{m+1,m+1})=0$. Ceci contredit rang $(M_{m+1,n}) = m+1$, donc rang $(M_{m+1,n}) = m$. Soit alors $(\delta_o,\dots,\delta_m)$ la relation décrite dans l'énoncé entre les lignes de $M_{m+1,n}$. Considérons la matrice $(m-n+1)\times(n+1)$:

$$N = \begin{pmatrix} \delta_o, & \delta_1, & \dots, & \delta_m, & 0, & \dots, & 0 \\ 0, & \delta_o, & \dots, & & \delta_m, & 0, \dots, & 0 \\ \vdots & & \ddots & & & \ddots & \\ 0, & & 0, & \delta_o, & \dots & \dots, & \delta_m \end{pmatrix}.$$

On vérifie immédiatement l'égalité $M_{m,n+1} \cdot {}^tN = 0$. D'après ([9] n° 230, p 231, ou [2] § 8, ex 13), il existe des polynômes homogènes en $(X_o,\dots,X_{r-1})$, premiers entre eux, Q et R tels que pour toute partie I à m éléments de $[0,n]$ on a :

$$Q \cdot D_I(M_{m,n+1}) = \pm R \cdot D_{CI}(N) ,$$

où $D_I(M_{m,n+1})$ (resp. $D_{CI}(N)$) désigne le mineur maximal de $M_{m,n+1}$ (resp. N) porté par les colonnes d'indices situés dans I (resp. CI le complémentaire de I dans $[0,n]$) . Il est clair que les mineurs maximaux de N engendrent l'idéal $(\delta_o,\dots,\delta_m)^{n-m+1}$, donc que le p.g.c.d. de ces mineurs est 1 ce qui entraîne que Q est une constante non nulle. Comme R est non nul, l'assertion est démontrée compte tenu de 1.1.

Démontrons maintenant le théorème. Sous les hypothèses du lemme précédent, montrons que V est contenue dans une hypersurface de degré $\leq e+r-n+\frac{P(n+1)}{n+1-m}$. Soit $(\delta_o,\ldots,\delta_m)$ la relation donnée dans le lemme entre les lignes de $M_{m+1,n}$. L'hypersurface d'équation $\delta_o+\delta_1 X_r+\ldots+\delta_m X_r^m$ contient V car

$$u_o(\delta_o+\ldots+\delta_m X_r^m) = \delta_1(u_o X_r - u_1)+\ldots+\delta_m(u_o X_r^m - u_m) .$$

Cette hypersurface a pour degré $m+d^o(\delta_m) \leq m+\frac{P(m)}{n-m+1}$ (en vertu du lemme). En simplifiant, on trouve $s \leq e+r-n+\frac{P(n+1)}{n+1-m}$.

Supposons d'abord qu'il existe un entier t tel que $P(t)<0$; soit alors n maximal pour cette propriété. Comme $\text{dét}(M_{n,n})/F^{n-1}$ est de degré $P(n)$, on a $\text{dét}(M_{n,n}) = 0$. Soit m un entier maximal tel que $m<n$ et que $\text{dét}(M_{m,m}) \neq 0$. On a alors $s \leq e+r-n+\frac{P(n+1)}{n+1-m}$. Mais $P(n+1) < P(n+1)-P(n) = 2n+1-(e+r+1) < n+1-m$ (cette dernière inégalité étant vérifiée car n est strictement contenu entre les racines de P et m est inférieur ou égal à la plus petite racine). On en déduit $s \leq e+r-n$ ce qui entraîne que s est $\leq$ à la plus petite racine de P, et en particulier $s \leq [P(l)]_+ + l$ pour tout entier l.

S'il n'existe pas d'entier t tel que $P(t) \leq 0$, soit n un entier tel que $n^2-(e+r)n+d = P(n)+n$ soit minimum. Si $\text{dét}(M_{n,n}) \neq 0$, soit $(\delta_o,\ldots,\delta_n)$ la relation décrite dans le lemme entre les lignes de $M_{n+1,n}$. L'hypersurface d'équation $\delta_o+\delta_1 X_r+\ldots+\delta_n X_r^n$ contient V. Comme $d^o\delta_n \leq d^o(\text{dét}\, M_{n,n}/F^{n-1}) = P(n)$, c'est terminé. Si $\text{dét}(M_{n,n}) = 0$, considérons comme précédemment le plus grand entier m tel que $m<n$ et $\det(M_{m,m}) \neq 0$. On sait que $s \leq e+r-n+\frac{P(n+1)}{n+1-m}$. Il suffit alors de remarquer que le terme de droite de cette inégalité croît avec m et que sa valeur pour $m = n$

est $n+P(n)$.

Il ressort de ce que nous venons de démontrer que s est toujours $\leq$ à la plus petite racine de P lorsqu'il y en a une. On en déduit que pour $P(n) \leq 0$ l'égalité $s = [P(n)]_+ + n$ est réalisée si et seulement si $s = n$ est la plus petite racine de P . Montrons que cette propriété caractérise les intersections complètes d'hypersurfaces de degrés respectifs s et $d/s = e+r+1-s$.

Soit S une hypersurface de degré s contenant V , et soit J l'idéal de V sur S . La suite exacte

$$0 \longrightarrow O_S \longrightarrow J^V \longrightarrow \omega_V(r+1-s) \longrightarrow 0 \qquad (*)$$

montre que $J^V(s-r-1-e)$ a une section de degré $d-s(e+r+1-s) = P(s)$. Il est immédiat que $P(s) = 0$ si et seulement si $J \xrightarrow{\sim} O_S(e+r+1-s)$, et le théorème est demontré.

Remarque 1.4: La suite exacte (*) ci dessus, et la conséquence qu'on en tire - $P(s) \geq 0$, l'égalité ayant lieu si et seulement si V est intersection complète d'hypersurfaces de degrés s et d/s - sont valables pour toute variété V (éventuellement non réduite) purement de codimension 2, contenue dans une hypersurface intègre de degré s . On retrouve là un résultat bien connu (voir par exemple [10], vanishing Lemma D, p.122).

Remarque 1.5: Dans le chapitre 3 de son mémoire, Halphen considère ce théorème comme évident pour les courbes de $\mathbb{P}^3$, car il affirme $\det(M_{i,i}) \neq 0$ pour tout $i \leq s$, lorsque la projection $\overline{C}$ de la courbe n' a que des points doubles ordinaires. Nous ne savons pas si cette assertion est vraie pour un centre de projection général, mais on constate facilement qu'elle est fausse pour un centre de projection particulier.

Exemple: Soit C l'intersection de deux surfaces quartiques générales de $\mathbb{P}^3$, et soit O n'appartenant pas à C un point double sur l'unique surface quartique contenant C et passant par O . Alors la projection de C , de centre O , n'a que des points doubles ordinaires et $\text{dét}(M_{3,3}) = 0$.

Toutefois le chapitre 2 du même mémoire contient une démonstration correcte du théorème de spécialité lorsque $P(3) < 0$.

Remarque 1.6: Lorsque V est section d'un fibré E de rang 2 sur $\mathbb{P}^r$ (i.e. lorsque $O_V(e)$ est un module dualisant sur V), la partie du théorème qui traite du cas où P a des racines est une conséquence immédiate d'un résultat de Barth ([1], cor.1 du Th.3). On a en effet $c_1(E) = e + r + 1$ et $c_2(E) = d$, donc P a des racines si et seulement si le discriminant $c_1^2 - 4c_2$ de E est positif. Le résultat signifie alors que E est instable donc que $2s \leq c_1$ ce qui démontre le théorème compte tenu de $P(s) \geq 0$.

§ 2. Courbes arithmétiquement de Cohen-Macaulay dans l'espace.

La classification numérique complète des courbes arithmétiquement normales de $\mathbb{P}^3$ repose essentiellement sur le résultat qui suit, et sur le théorème 2.5.

Proposition 2.1: Soit V une variété intègre de $\mathbb{P}_k^r = \text{Proj}(k[X_o, \ldots, X_r])$ et soit M un O_V-module cohérent sans torsion. On suppose que $O = (0,0,\ldots,0,1)$ n'appartient pas à V . Soit alors p la projection de V sur le plan à l'infini $\mathbb{P}_k^{r-1}$. Soit m_i $(i \geq 0)$ la suite rangée par ordre croissant des degrés d'un système de générateurs du $k[X_o, \ldots, X_{r-1}]$-module $\sum_{n \in Z} H^o(M(n))$. Si k est un entier positif tel que $m_{k-1} + 1 < m_k$, alors il existe une hypersurface de degré k de $\mathbb{P}^r$ contenant V , et ne contenant pas O .

Si a_i est le générateur de degré m_i , il existe des éléments $\alpha_{i,j} \in k[X_o,\ldots,X_{r-1}]_{i-j+1}$ avec $0 \leq i,j \leq k-1$ tel que $X_r a_i = \sum_o^{k-1} \alpha_{ij} a_j$. L'hypersurface d'équation $\text{dét}(\alpha_{ij} - \delta_{ij} X_r)$ a la propriété annoncée.

<u>Corollaire 2.2: Soit V une variété intègre de codimension 2 de $\mathbb{P}^r$, projectivement de Cohen-Macaulay. Soit s une hypersurface de degré minimum s contenant V . Supposon $0 \notin s$; alors une résolution minimale de $\sum_{l \geq o} H^o(\mathcal{O}_V(l))$ sur $k[X_o,\ldots,X_{r-1}]$ induit une suite exacte</u>

$$0 \longrightarrow \bigoplus_o^{s-1} \mathcal{O}_{\mathbb{P}^{r-1}}(-n_i) \xrightarrow{\varphi} \bigoplus_o^{s-1} \mathcal{O}_{\mathbb{P}^{r-1}}(-i) \longrightarrow \mathcal{O}_V \longrightarrow 0$$

<u>où</u> $n_{i-1} \geq n_i \geq n_{i-1} - 1 \quad (1 \leq i \leq s-1)$ <u>et</u> $n_{s-1} \geq s$.

Soit $k[x_o,\ldots,x_r]$ le cône projetant de V . Il est de dimension projective 1 sur $k[X_o,\ldots,X_{r-1}]$, et admet comme système minimal de générateurs les éléments $1, x_r, \ldots, x_r^{s-1}$. Comme V n'est pas sur une hypersurface de degré $< s$, on a $n_i \geq s$ pour tout i . L'absence de trou dans la suite, que nous avons rangée par ordre décroissant, des n_i se déduit du fait que la transposée de φ est une présentation du $\mathcal{O}_V$-module sans torsion $\underline{\text{Ext}}^1_{\mathcal{O}_{\mathbb{P}^{r-1}}}(\mathcal{O}_V, \mathcal{O}_{\mathbb{P}^{r-1}})$.

Remarque 2.3: On en déduit immédiatement le théorème de spécialité pour les variétés intègres de codimension 2 dans $\mathbb{P}^r$, projectivement de Cohen-Macaulay. En effet, on a $e + r = n_o$. Comme on peut supposer V birationnelle avec son image dans $\mathbb{P}^{r-1}$, si d est le degré de V , on a $d = d^o(\det(\varphi)) = \sum_o^{s-1} (n_i - i)$. On en déduit pour tout l $(1 \leq l \leq s)$

$$d \geq \sum_o^{l-1} (n_i - i) \geq \sum_o^{l-1} (n_o - 2i) \geq l n_o - l(l-1) ,$$

c'est à dire $d \geq l(e+r+1-l)$. On remarque de plus que l'égalité n'est réalisée que lorsque $l = s$ et $n_i = n_o - i$ pour tout i , ce qui caractérise les intersections complètes de surfaces de degrés respectifs s et $n_o + 1 - s$.

Definition 2.4: Soit V une variété projectivement de Cohen-Macaulay, de codimension 2 dans $\mathbb{P}^r$. Supposons V contenue dans une hypersurface de degré s de $\mathbb{P}^r$ et non contenue dans une hypersurface de degré $< s$. Une projection suffisamment générale de V sur l'hyperplan à l'infini $\mathbb{P}^{r-1}$ induit une suite exacte

$$0 \longrightarrow \bigoplus_{o}^{s-1} O_{\mathbb{P}^{r-1}}(-n_i) \longrightarrow \bigoplus_{o}^{s-1} O_{\mathbb{P}^{r-1}}(-i) \longrightarrow O_V \longrightarrow 0 ,$$

avec $n_o \geq n_1 \geq \dots \geq n_{s-1} \geq s$. Nous dirons que la suite $(n_o,\dots,n_{s-1})$ est le <u>caractère numérique</u> de V . Il est évident que le caractère numérique de V ne dépend que de la fonction de Hilbert de V et la caractérise.

Nous avons vu que si V est intègre, son caractère numérique est sans lacune, i.e. $n_i \leq n_{i+1} + 1$ pour $i \geq 0$. Montrons inversement que tout caractère numérique sans lacune est effectif pour une courbe arithmétiquement normale de $\mathbb{P}^3$.

<u>Théorème 2.5</u>. <u>Soit</u> $(n_i)_{0\leq i\leq s-1}$ <u>une suite décroissante d'entiers tels que</u> $n_i \leq n_{i+1} + 1$ $(0 \leq i \leq s-2)$ <u>et</u> $s \leq n_{s-1}$. <u>Il existe une courbe arithmétiquement normale de</u> $\mathbb{P}^3$ <u>de caractère numérique</u> $(n_i)_{0\leq i\leq s-1}$.

Rappelons que pour toute courbe C de $\mathbb{P}^3$, nous appelons $e(C)$ le plus grand entier n tel que $H^o(\omega_C(-n)) \neq 0$. Nous allons montrer par récurrence sur s que toute suite $(n_i)_{0\leq i\leq s-1}$ ayant les propriétés énoncées peut être réalisée comme caractère numérique d'une courbe arithmétiquement normal C de $\mathbb{P}^3$ vérifiant la condition

(*) $\omega_C(-e(C))$ a une section sans point multiple.

Soit donc C une courbe arithmétiquement normale vérifiant (*) de caractère numérique $(n_o,\ldots,n_{s-1})$; nous devons prouver l'existence d'une courbe arithmétiquement normale C_1 (resp. C_2), vérifiant (*) et de caractère numérique $(n_o+2, n_o+1, n_1+1,\ldots, n_{s-1}+1)$ (resp. $(n_o+1, n_o+1, n_1+1,\ldots, n_{s-1}+1)$). La résolution

$$0 \longrightarrow \bigoplus_{o}^{s-1} O_{\mathbb{P}^2}(-n_i) \longrightarrow \bigoplus^{s-1} O_{\mathbb{P}^2}(-i) \longrightarrow O_C \longrightarrow 0$$

prouve d'une part que $n_o = e(C) + 3$, et d'autre part que C est intersection de surfaces de $\mathbb{P}^3$ de degrés $\leq n_o$. Il existe donc une surface lisse F contenant C de degré n_o+2 (resp. n_o+1). Soit L le faisceau inversible sur F dont C est une section. Nous allons montrer qu'on peut prendre pour C_1 (resp. C_2) une section suffisamment générale de $L(1)$. La suite exacte $0 \to L^{-1} \to O_F \to O_C \to 0$ induit une suite exacte

$$0 \longrightarrow O_F \longrightarrow L \longrightarrow \omega_C(4-(n_o+2)) \longrightarrow 0$$

$$(\text{resp. } 0 \longrightarrow O_F \longrightarrow L \longrightarrow \omega_C(4-(n_o+1)) \longrightarrow 0)$$

Relevons une section lisse de $\omega_C(-(n_o-3))$ (resp. $\omega_C(-(n_o-4))$) en une section lisse C_1 (resp. C_2) de $L(1)$. Comme $H^1(L^{-1}(n)) = 0$ pour tout n, la courbe C_1 (resp. C_2) est arithmétiquement normale. La suite exacte

$$0 \longrightarrow O_F \longrightarrow L(1) \longrightarrow \omega_{C_1}(4-(n_o+2)) \longrightarrow 0$$

$$(\text{resp. } 0 \longrightarrow O_F \longrightarrow L(1) \longrightarrow \omega_{C_2}(4-(n_o+1)) \longrightarrow 0)$$

démontre $e_1 = e(C_1) = e(C)+2$ (resp. $e_2 = e(C_2) = e(C)+1$). La section de $L(1)$ étant un rélèvement d'une section lisse de $\omega_C(-e)$ (resp. $\omega_C(-e+1)$), le diagramme qui suit montre que $\omega_{C_1}(-e_1)$

(resp. $\omega_{C_2}(-e_2)$) a une section lisse, de fait la même, c'est l'intersection schématique de C et C_1 (resp. C et C_2) .

$$\begin{array}{ccccccccc}
 & & & & 0 & & & & \\
 & & & & \downarrow & & & & \\
 & & & & O_F & = & O_F & & \\
 & & & & \downarrow & & \downarrow & & \\
0 & \longrightarrow & O_F(1) & \longrightarrow & L(1) & \longrightarrow & \omega_C(-e) & \longrightarrow & 0 \\
 & & \| & & \downarrow & & \downarrow & & \\
 & & O_F(1) & \longrightarrow & \omega_{C_1}(-e+1) & \longrightarrow & \text{coker} & \longrightarrow & 0 \\
 & & & & \downarrow & & \downarrow & & \\
 & & & & 0 & & 0 & &
\end{array}$$

$$\text{(resp.}\quad \begin{array}{ccccccc}
 & & O_F & = & O_F & & \\
 & & \downarrow & & \downarrow & & \\
O_F(1) & \longrightarrow & L(1) & \longrightarrow & \omega_C(-e+1) & \longrightarrow & 0 \\
\| & & \downarrow & & \downarrow & & \\
O_F(1) & \longrightarrow & \omega_{C_2}(-e_2+1) & \longrightarrow & \text{coker} & \longrightarrow & 0 \\
 & & \downarrow & & \downarrow & & \\
 & & 0 & & 0 & &
\end{array}\quad)$$

Il reste à démontrer que C_1 et C_2 ont les caractères numériques annoncés. Faisons-le pour C_1, la démonstration pour C_2 étant identique. La suite exacte $0 \to L^{-1} \to O_F \to O_C \to 0$ montre que s est le plus petit entier n tel que $L^{-1}(n)$ ait une section. De la suite exacte $0 \to L^{-1}(-1) \to O_F \to O_{C_1} \to 0$, on déduit que C_1 est contenue dans une surface de degré $s+1$ et non dans une surface de degré moindre. Puisque $O_F = \bigoplus^{n_o+1} O_{\mathbb{P}^2}(-i)$, le diagramme commutatif suivant montre que $L^{-1} \simeq \bigoplus^{s-1} O_{\mathbb{P}^2}(-n_i) \oplus \bigoplus_s^{n_o+1} O_{\mathbb{P}^2}(-i)$:

$$\begin{array}{ccccccc}
0 \longrightarrow L^{-1} & \longrightarrow & \bigoplus_{o}^{n_o+1} \mathcal{O}_{\mathbb{P}^2}(-i) & \longrightarrow & \mathcal{O}_C & \longrightarrow & 0 \\
\downarrow & & \downarrow & & \| & & \\
0 \longrightarrow \bigoplus_{o}^{s-1} \mathcal{O}_{\mathbb{P}^2}(-n_i) & \longrightarrow & \bigoplus_{o}^{s-1} \mathcal{O}_{\mathbb{P}^2}(-i) & \longrightarrow & \mathcal{O}_C & \longrightarrow & 0 .
\end{array}$$

Si $(m_o,\ldots,m_s)$ est le caractère numérique de C_1, le diagramme commutatif suivant démontre le résultat annoncé

$$\begin{array}{ccccccc}
0 \longrightarrow L^{-1}(-1) & \longrightarrow & \bigoplus_{o}^{n_o+1} \mathcal{O}_{\mathbb{P}^2}(-i) & \longrightarrow & \mathcal{O}_{C_1} & \longrightarrow & 0 \\
\downarrow & & \downarrow & & & & \\
0 \longrightarrow \bigoplus_{o}^{s} \mathcal{O}_{\mathbb{P}^2}(-m_i) & \longrightarrow & \bigoplus_{o}^{s} \mathcal{O}_{\mathbb{P}^2}(-i) & \longrightarrow & \mathcal{O}_{C_1} & \longrightarrow & 0 .
\end{array}$$

Remarque 2.6: Si C est une courbe arithmétiquement de Cohen-Macaulay de $\mathbb{P}^3$, de caractère numérique $(n_o,\ldots,n_{s-1})$, de degré d et de genre arithmétique g, on a

$$d = \sum_{o}^{s-1} (n_i - i) \quad \text{et} \quad g = 1 + 1/2 \left[\sum_{o}^{s-1} (n_i - i)(n_i + i - 3) \right] .$$

Théorème 2.7: Soient d et s des entiers positifs tels que $d \geq s(s+1)/2$. Soit $(m_o,\ldots,m_{s-1})$ la suite maximale pour l'ordre lexicographique parmi les suites $(n_o,\ldots,n_{s-1})$ vérifiant

1) $n_o \geq n_1 \geq \ldots \geq n_{s-1} \geq s$.

2) $n_i \leq n_{i+1} + 1$ pour $0 \leq i \leq s-2$.

3) $\sum_{o}^{s-1} (n_i - i) = d$.

A) Soit Δ une courbe arithmétiquement normale de $\mathbb{P}^3$ admettant le caractère numérique $(m_o,\ldots,m_{s-1})$. Alors pour toute courbe intègre C de $\mathbb{P}^3$, arithmétiquement de Cohen-Macaulay, de degré d, non contenue dans une surface de degré $< s$, on a

<u>$h^1(O_C(n)) \leq h^1(O_\Delta(n))$ pour tout n ; de plus si C ne présente pas le même caractère numérique que Δ on a $h^1(O_C(l)) < h^1(O_\Delta(l))$ pour $l \leq s$.</u>

B) <u>Si $s(s-1) < d$, posons $d + r = st$ avec $0 \leq r < s$. Alors Δ est liée a une courbe plane de degré r par deux surfaces de degrés s et t . On a</u>

<u>$e(\Delta) = s + [d/s] - 4$ et</u>

<u>$g(\Delta) = 1 + d/2(s + d/s - 4) - r(s-r)(s-1)/2s$.</u>

<u>Si $s(s-1) \geq d$, considérons les entiers ν et μ tels que $0 \leq \nu \leq \mu \leq s-3$ et et</u>

<u>$d = s(s+1)/2 + \mu(\mu+1)/2 + \nu$. On a alors</u>

<u>$e(\Delta) = s + \mu - 3$ et</u>

<u>$g(\Delta) = 1 + s(s+1)(2s-5)/6 + \mu(\mu+1)(\mu+3s-4)/6 + \nu(\nu+2s-3)/2$.</u>

D'après le théorème précédent, il existe bien une telle courbe Δ Notons Φ son caractère numérique. Pour tout $\sigma \geq s$, et tout caractère sans lacune $\chi = (n_o,\ldots,n_{\sigma-1})$ vérifiant $n_o \geq n_1 \geq \ldots \geq n_{\sigma-1}$ et $\sum_o^{\sigma-1} (n_i - i) = d$, considérons la fonction

$$h^1_\chi(n) = \left[\sum_{i=o}^{\sigma-1} (n_i-n-1)_+ - \sum_{i=o}^{\sigma-1} (i-n-1)_+\right] .$$

Cette fonction s'interprète de la façon suivante: Si Γ est un groupe de points de $\mathbb{P}^2$ de caractère numérique χ , soit J_Γ l'idéal de Γ dans $\mathbb{P}^2$, alors $h^1(J_\Gamma(n)) = h^1_\chi(n)$. Si C est une courbe arithmétiquement Cohen-Macaulay de $\mathbb{P}^3$ de caractère numérique χ , on a $h^1(O_C(n)) = \sum_{l \geq n+1} h^1_\chi(l)$. Pour démontrer la première partie de A), il suffit donc de montrer $h^1_\chi(n) \leq h^1_\Phi(n)$ pour $n \geq 0$. Introduisons la fonction en escalier suivante sur $\mathbb{R}^+$:

$F_\chi(x) = [x]+1$ pour $x<\sigma$, et

$F_\chi(x)$ = nombre des n_i vérifiant $n_i \geq x$ pour $x \geq \sigma$.

Remarquons que $d = \sum_0^{\sigma-1} (n_i - i) = \int_0^\infty F_\chi(x)dx$.

L'inégalité $h^1_\chi(n) \leq h^1_\Phi(n)$ s'écrit alors $\int_{n+1}^\infty [F_\Phi(x) - F_\chi(x)]dx \geq 0$.

Comme Φ est maximum pour l'ordre lexicographique $n_p < m_p$ entraîne $n_q \leq m_q$ pour $q \leq p$. On en déduit que la fonction $[F_\Phi - F_\chi]$ est d'abord négative et ensuite positive. Comme $\int_0^\infty [F_\Phi(x) - F_\chi(x)]dx = 0$, **la valeur de** $\int_n^\infty [F_\Phi(x) - F_\chi(x)]dx$ est bien toujours positive.

De plus, si $\chi \neq \Phi$, il existe p tel que $m_p > n_p \geq s+1$. On en déduit $n_q \leq m_q$ pour $q \leq p$ et $\int_{n_p+1}^\infty [F_\Phi(x) - F_\chi(x)]dx > 0$, soit $h^1_\chi(n_p) < h^1_\Phi(n_p)$ ce qui entraîne bien $h^1(O_C(l)) < h^1(O_\Delta(l))$ pour $l \leq s$.

Pour démontrer B), exhibons le caractère numérique Φ . Lorsque $s(s-1) < d$, on a

$$\Phi = (s+t-2, s+t-3, \ldots, s+t-r-1, s+t-r-1, \ldots, t) \quad \text{si} \quad r \neq 0$$

$$\Phi = (s+t-1, s+t-2, \ldots, t) \quad \text{si} \quad r = 0 .$$

La courbe Δ est alors située sur une surface T de degré t ne contenant pas la surface S de degré s . Considérons la courbe Γ liée à Δ par S et T . La suite exacte $0 \to \omega_\Delta(4-s-t) \to O_{S\cap T} \to O_\Gamma \to 0$ montre que Γ est plane car $e(\Delta) = s + [d/s] - 4$. La formule donnant le genre g de Δ s'obtient soit par calcul direct soit en appliquant ([12], prop.3.1). Ce résultat est signalé par Halphen ([6], p.401).

Lorsque $s(s-1) \geq d$, on a

$$\Phi = (s+\mu, s+\mu-1, \ldots, s+\nu, s+\nu, s+\nu-1, \ldots, s, s, \ldots, s) .$$

On a donc bien $e(\Delta) = s + \mu - 3$, l'expression du genre se déduit d'un calcul direct mais fastidieux. Le théorème est demontré.

Remarque 2.8: Le meilleur moyen de se convaincre que les caractères exhibés sont lexicographiquement maximaux est de tracer les graphes des fonctions F_{Φ} introduites dans la démonstration de A).

Remarque 2.9: Le lecteur a constaté que pour $s(s-1) \geq d$ nous n'avons pas décrit de courbe Δ admettant le caractère lexicographiquement maximal. Lorsque $d = s(s+1)/2$ le seul caractère numérique possible est $n_i = s$ pour $0 \leq i \leq s-1$. La courbe correspondante Δ est intersection des surfaces dont les équations sont les mineurs maximaux d'une matrice $(s+1) \times s$ à coefficients linéaires suffisamment généraux dans $k[X_0,\ldots X_3]$. On peut la construire aussi par récurrence sur s, comme dans la démonstration du théorème 2.5.

Pour tout (s,μ,ν) tel que $s-3 \geq \mu \geq \nu$, indiquons une construction géométrique pour trouver une courbe $\Delta(s,\mu,\nu)$. On pose $\lambda = s-\mu-2$ On considère d'abord une courbe $\Delta(\lambda-1,0,0)$ (décrite plus haut); convenons que $\Delta(0,0,0)$ est vide. Soit L une droite ne rencontrant pas $\Delta(\lambda-1,00)$. On peut trouver une courbe lisse C_1, liée à $\Delta(\lambda-1,0,0)$ par des surfaces de degrés λ et $(s-1)$, et admettant L comme λ-sécante. On démontre alors que la courbe réduite $C_1 \cup L$ est arithmétiquement Cohen-Macaulay et qu'elle est liée par deux surfaces générales de degrés s à une courbe $\Delta(s,\nu,\nu)$. Pour construire une courbe $\Delta(s,\mu,\nu)$, on considère une courbe $\Delta(s-\mu+\nu,\nu,\nu)$ (dont on vient de voir la construction), qu'on place sur une surface lisse de degré $(s+\nu+1)$. Si $\Delta(s-\mu+\nu,\nu,\nu)$ est une section du faisceau inversible L sur cette surface, alors $\Delta(s,\mu,\nu)$ est une section de $L(\mu-\nu)$. On remarque que cette dernière partie de la construction est la même (en remplaçant 1 par $\mu-\nu$) que la construction de la courbe C_1 dans la démonstration de 2.5; on procède donc de la même façon pour obtenir une courbe lisse. Nous ne savons pas si cette construction permet d'obtenir la courbe $\Delta(s,\mu,\nu)$ la plus générale, sauf pour $\mu = \nu = 0$ où c'est le cas, et pour $\nu = 1$ où nous sommes sûrs du contraire.

Bien que cela ne soit pas expressément dit, il résulte immediatement de ([4], th.2) que deux courbes arithmétiquement Cohen-Macaulay de $\mathbb{P}^3$ appartiennent à la même composante irréductible du schéma de Hilbert si et seulement si elles ont même fonction de Hilbert. En effet, à toute courbe arithmétiquement Cohen-Macaulay de $\mathbb{P}^3$, Ellingsrud associe son type $(n_{ij}(C))$ donné par une résolution minimale sur $k[X_0,\dots,X_3]$ du cône projetant de C dans $\mathbb{P}^3$; il remarque que la composante irréductible de C dans le schéma de Hilbert est définie par son type réduit (minimal pour la relation de prolongation). On voit facilement que le type réduit caractérise la fonction de Hilbert de C. On en déduit le résultat qui suit:

<u>Proposition 2.10</u>: <u>Deux courbes arithmétiquement de Cohen-Macaulay de $\mathbb{P}^3$ sont dans la même composante irréductible du schéma de Hilbert si et seulement si elles ont même caractère numérique</u>.

Compte tenu de 2.5, on obtient l'énoncé suivant:

<u>Théorème 2.11</u>: <u>Toute courbe intègre arithmétiquement Cohen-Macaulay de $\mathbb{P}^3$ admet une générisation arithmétiquement normale (donc lisse) dans le schéma de Hilbert</u>.

Remarquons la relation suivante entre le type réduit et le caractère numérique χ d'une courbe arithmétiquement Cohen-Macaulay de $\mathbb{P}^3$. Si F_χ est la fonction en escalier définie dans la démonstration du théorème 2.7, on a:

$$[F_\chi(k+1) - 2F_\chi(k) + F_\chi(k-1)]_+ = \text{nombre de } n_{2j} \text{ égaux à } k .$$

$$[F_\chi(k+1) - 2F_\chi(k) + F_\chi(k-1)]_- = \text{nombre } n_{1j} \text{ égaux à } k, \text{ pour } k > s$$

$$F_\chi(s) - F_\chi(s+1) + 1 = \text{nombre de } n_{1j} \text{ égaux à } s .$$

Signalons enfin que pour toute courbe intègre arithmétiquement Cohen-Macaulay de $\mathbb{P}^3$, il existe une courbe arithmétiquement normale de $\mathbb{P}^3$ admettant pour type le type réduit de cette courbe. On le démontre en remarquant que dans la preuve de 2.5, si la courbe C a un type réduit alors pour un choix assez général de la section de $L(1)$ la courbe C_1 (resp. C_2) a un type réduit.

Pour conclure ce paragraphe donnons à titre d'exemple la classification (par leur caractère numérique) des courbes arithmétiquement normales de degré 20 de $\mathbb{P}^3$. La condition $s(s+1)/2 \leq d$ impose $s \leq$

1) Pour $s = 5$.

a) caractère numérique maximal $\Phi_0 = (7,7,6,5,5)$, $\mu = \nu = 2$. Une telle courbe est liée par des surfaces de degrés 5 et 7 à la courbe résiduelle d'une droite par deux surfaces quartiques.

b) $\Phi_1 = (7,6,6,6,5)$. Liée par deux surfaces quintiques à la courbe résiduelle d'une droite par une quadrique et une surface cubique

c) $\Phi_2 = (6,6,6,6,6)$. Liée par des surfaces de degrés 5 et 6 à une courbe de degré 10 de caractère numérique $(4,4,4,4)$

2) Pour $s = 4$.

a) Le caractère numérique maximal est $\chi_0 = (8,7,6,5)$. Une courbe présentant ce caractère est intersection complète de surfaces de degrés 4 et 5 .

b) $\chi_1 = (7,7,6,6)$. Liée a une biquadratique gauche par des surfaces de degrés 4 et 6 .

3) Pour $s = 3$. L'unique caractère numérique possible est celui d'une courbe liée à une droite par des surfaces de degrés 3 et 7 ; soit $\chi = (8,8,7)$.

4) Pour $s = 2$. La seule courbe arithmétiquement normale est intersection complète d'une quadrique et d'une surface de degré 10, elle présente le caractère numérique (11,10) .

§ 3. Majoration du genre d'une courbe lisse connexe de $\mathbb{P}^3$.

On se propose de majorer

$G(d,s) = \max\{$genre de C , où C est une courbe lisse connexe de degré d dans $\mathbb{P}^3$, non contenue dans une surface de degré $< s\}$

Dans le paragraphe précédent, on a determiné

$G_{C.M}(d,s) = \max\{$genre de C , où C est une courbe arithmétiquement normale de degré d dans $\mathbb{P}^3$, non contenue dans une surface de degré $< s\}$

Théorème 3.1: Si $s(s-1) < d$, on a

$$G(d,s) = G_{C.M}(d,s) = 1 + d/2(s+d/s-4) - r(s-r)(s-1)/2s ,$$

où $d + r \equiv 0$ (modulo s) et $0 \leq r < s$. Les courbes réalisant ce maximum sont les courbes lisses liées à une courbe plane de degré r par des surfaces de degrés s et $(d+r)/s$.

Suivant la méthode de Castelnuovo ([3]), nous étudierons la postulation d'une section plane assez générale d'une courbe pour majorer le genre de cette courbe.

Lemme 3.2: Soit C une courbe intègre de $\mathbb{P}^3$. Si H est un plan assez général de $\mathbb{P}^3$, le groupe de points $C \cap H$ de H présente un caractère numérique sans lacune.

Soit $\mathbb{P}^{3\vee}$ l'espace des plans de $\mathbb{P}^3$, et soit $H \subset \mathbb{P}^3 \times \mathbb{P}^{3\vee}$ le plan universel de $\mathbb{P}^3$. Alors $\Gamma = C \times \mathbb{P}^{3\vee} \cap H$ est intègre. Soit η le point générique de $\mathbb{P}_3^\vee$. Si J_η est l'idéal dans $\mathbb{P}^3_{k(\eta)}$ de $\Gamma \times k(\eta)$, pour un point fermé x suffisemment général de $\mathbb{P}^{3\vee}$ l'idéal J_x de $\Gamma \times k(x)$ dans $\mathbb{P}^3_{k(x)}$ aura la même fonction de Hilbert que J_η. Comme $\Gamma \times k(\eta)$ est une variété intègre de codimension 2 dans $H \times k(\eta)$ son caractère numérique et sans lacune; donc celui de $\Gamma \times k(x)$ dans $H \times k(x)$ aussi puisque c'est le même.

Remarque 3.3: Compte tenu du théorème 2.5 et de ce lemme, on voit que pour toute courbe intègre C de $\mathbb{P}^3$ une section plane assez générale de C a même postulation que la section plane d'une courbe arithmétiquement normale de même degré.

<u>Proposition 3.4: Soit C une courbe intègre de $\mathbb{P}^3$, de degré d et de genre g, non contenue dans une surface de degré $< s$. Soit σ le plus petit degré des courbes d'un plan général H contenant $C \cap H$. On a</u>

$$g \leq G_{C.M}(d,\sigma) - \binom{s-\sigma+2}{3} .$$

Soit Δ une courbe arithmétiquement normale non contenue dans une surface de degré $< \sigma$, et de genre $G_{C.M}(d,\sigma)$. Soit H un plan général, et soit $J_{C\cap H}$ (resp. $J_{\Delta\cap H}$) l'idéal du groupe de points $C \cap H$ (resp. $\Delta \cap H$) dans H. Les suites exactes (pour tout l)

$$H^1(J_{C\cap H}(l)) \longrightarrow H^1(O_C(l-1)) \longrightarrow H^1(O_C(l)) \longrightarrow 0$$

$$0 \longrightarrow H^1(J_{\Delta\cap H}(l)) \longrightarrow H^1(O_\Delta(l-1)) \longrightarrow H^1(O_\Delta(l)) \longrightarrow 0 ,$$

montrent $h^1(O_C(k)) \leq h^1(O_\Delta(k))$ pour tout $k \geq 0$, compte tenu de 3.3 et du caractère maximal de Δ. Appliquant le théorème de Riemann-Roch, on en déduit

$$g \leq G_{C.M}(d,\sigma) + h^0(O_\Delta(k)) - h^0(O_C(k)) \quad \text{pour tout } k .$$

L'inégalité de l'énoncé est une conséquence immédiate de celle-ci en prenant $k = s-1$.

La première partie du théorème se déduit alors d'un résultat de Laudal ([8], cor.2 du "generalized tri-secant lemma"):

Soit C une courbe lisse connexe de degré d non contenue dans une surface de degré $< s$, avec $s(s-1) < d$, alors pour un plan général H, le groupe de points $H \cap C$ n'est pas contenu dans une courbe de degré $< s$.

Pour la deuxième partie, il reste à démontrer le résultat bien connu suivant (compte tenu de 2.7.A)):

<u>Lemme 3.5:</u> <u>Soit C une courbe intègre de $\mathbb{P}^3$. Soient H un plan ne contenant pas C et $J_{H\cap C}$ l'idéal du groupe de points $H \cap C$. Alors si $h^1(O_C) = \sum_{n\geq 1} h^1(J_{H\cap C}(n))$, C est arithmétiquement Cohen-Macaulay.</u>

Les suites exactes $H^1(J_{H\cap C}(n) \to H^1(O_C(n-1)) \to H^1(O_C(n)) \to 0$, pour $n \geq 1$ montrent que $h^1(O_C) = \sum_{n\geq 1} h^1(J_{H\cap C}(n))$ entraîne l'exactitude des suites

$H^1(J(n-1)) \to (H^1(J(n)) \to 0$, où J est l'idéal de C dans $\mathbb{P}^3$. Comme $H^1(J(0)) = 0$, on en déduit $H^1(J(n)) = 0$ pour tout n, et C est arithmétiquement de Cohen-Macaulay.

Remarque 3.6: Le théorème de spécialité est une conséquence immédiate du théorème 3.1, compte tenu de l'inégalité $ed \leq 2g-2$.

Remarque 3.7: La majoration du genre se démontre élémentairement lorsque la courbe C est contenue dans une surface de degré s, avec $s(s-1) \leq d$. Plus généralement si C est une courbe, éventuellement non réduite, de degré d, contenue dans une surface intègre S de degré s, on a

$$g(C) \leq 1 + d/2(s+d/s-4) - r(s-r)(s-1)/2s ,$$

où $d + r \equiv 0$ (modulo s) et $0 \leq r < s$.

Le résultat se démontre par restriction à un groupe de d point éventuellement multiples, sur une section plane intègre de S , par récurrence sur le nombre minimal de générateurs de l'idéal gradué du groupe de points, au moyen de techniques de liaison, en remarquant que l'expression de droite dans l'inégalité est une fonction croissante de s pour $s \geq \sqrt{d}$.

Terminons ce paragraphe en remarquant que la majoration du genre pour les courbes arithmétiquement normales de $\mathbb{P}^3$ n'est pas valable pour les courbes lisses connexes générales, plus précisement qu'on n'a pas $G(d,s) = G_{C.M}(d,s)$. Compte tenu des résultats précédents, pour démentir cette égalité on est amené à rechercher une courbe lisse connexe de $\mathbb{P}^3$, non contenue dans une surface de degré s et telle que toute section plane de cette courbe soit contenue dans une courbe de degré s , avec $s(s+1) \geq d$ et $\binom{s+2}{2} \leq d$. Voici un exemple de degré minimum suggéré par une remarque de Laudal: Prenons pour C une courbe de degré 10 et de genre 11 liée par deux surfaces quartiques générales à une courbe de degré 6 et de genre 3 tracée sur une quadrique. On vérifie facilement que C n'est pas sur une surface cubique. Si une section plane $C \cap H$ de C n'est pas sur une cubique de H , elle ne peut qu'avoir le caractère numérique (4,4,4,4 dans ce cas si $J_{H \cap C}$ est l'idéal dans H de $H \cap C$, on a $\sum_{n \geq 1} h^1(J_{H \cap C}(n)) = 11$. D'après 3.5, l'existence d'une telle section plane entraînerait que C est arithmétiquement normale ce qui est faux; donc toute section plane est sur une cubique. Cet exemple ne dément pas la majoration du genre. En fait, il ne faut pas aller chercher beaucoup plus loin un contre-exemple à cette majoration: En effet, si E est le fibré de rang 2 bien connu admettant pour section deux droites disjointes, la courbe de degré 10 et de genre 11 considérée plus haut est une section de $E(2)$; soit maintenant C

une section lisse nécessairement connexe de $E(3)$; cette courbe de degré 17 et de genre 35 n'est pas sur une surface quartique. Or d'après le théorème 2.7, une courbe arithmétiquement normale de degré 17, non contenue dans une quartique est de genre ≤ 34.

De façon plus générale, la méthode suivante permet de fabriquer des contre-exemples à cette majoration:

Proposition 3.8: Soit E un fibré de rang 2 sur $\mathbb{P}^3$, de classes de Chern c_1 et c_2. On suppose $H^0(E(-1)) = 0$, et $H^0(E) \neq 0$ et $c_1(c_1+1) > 2c_2$. Alors pour $s \gg 0$, une section générale de $E(s-c_1)$ est une courbe lisse connexe C telle que

1) $d(C) = s^2 - c_1 s + c_2$; $e(C) = 2s - c_1 - 4$
2) C est contenue dans une surface de degré s et non contenue dans une surface de degré $s-1$, et le genre de C vérifie l'inégalité $g(C) > G_{C.M.}(d,s)$.

Avant de démontrer ce résultat, remarquons que de tels fibrés existent. Signalons les exemples suivants:

a) Fibré admettant pour section deux droites disjointes. Dans ce cas $c_1 = c_2 = 2$, et l'assertion est vérifiée pour $s \geq 5$.

b) Fibré admettant pour section une courbe elliptique de degré 7 (resp. 8,9) contenue dans une surface quartique et non dans une surface cubique. Dans ce cas $c_1 = 4$ et $c_2 = 7$ (resp. 8,9); l'assertion est vérifiée pour $s \geq 8$.

c) Fibré admettant pour section une courbe canonique de genre 7 (resp. 8) contenue dans une surface quintique et non dans une surface quartique. Dans ce cas $c_1 = 5$ et $c_2 = 12$ (resp. 14).

d) Fibrés de discriminant arbitrairement grand en prenant l'imag réciproque de l'un des fibrés décrits plus haut, par un endomorphisme de $\mathbb{P}^3$ donné par 4 polynômes homogènes de même degré sans zéro commun.

Démontrons maintenant $g(C) > G_{C.M}(d,s)$ la seule assertion non évidente de l'énoncé. Si μ et ν sont les entiers vérifiant $0 \leq \nu \leq \mu \leq s-3$, et $d = s(s+1)/2 + \mu(\mu+1)/2 + \nu$, on voit immédiatement que $\mu \leq s - c_1 - 2$, et que pour $s \geq \binom{c_1+2}{2} - c_2$ on a $\mu = s - c_1 - 2$ et $\nu = s - \binom{c_1+2}{2} + c_2$. On trouve

$$g(C) - G_{C.M.}(d,s) = ks - (k+c_1+c_2+1)c_1/6 - k/2(k+c_1+4) ,$$

où $k = \binom{c_1+2}{2} - c_2$. Il est clair que cette valeur est positive pour $s >> 0$.

§ 4. Exemples.

Nous donnons dans ce paragraphe deux exemples de classification d courbes. Le choix de ces exemples n'est pas arbitraire. Dans les deux cas, on a $e \leq 1$, et dans les deux cas la condition $e = 1$ implique une diminution du degré minimum des surfaces contenant la courbe, ce phénomène n'étant pas expliqué par les résultats démontrés dans cet article.

La classification des courbes de degré 8 et genre 5 est complète. Son aspect est simple et reproduit fidèlement les sous-familles classiques qu'on distingue dans les modules de courbes. Pour $d = 10$ et $g = 6$, nous nous sommes contentés de décrire les types de grande généralité. Une analyse plus complète met en évidence des phénomènes qu'il est difficile d'unifier. Notons toutefois l'importance que peut avoir la présence de droites 6, 7 ou 8-sécantes.

Le schéma de Hilbert H_8^5 des courbes lisses connexes de degré 8 et de genre 5 de $\mathbb{P}^3 = P$ est intègre, normal, localement intersection complète, de dimension 32. On obtient les cinq localement fermés suivants de H_8^5 :

A) Une courbe générique présente une résolution

$$0 \longrightarrow O_P^2(-6) \longrightarrow O_P^8(-5) \longrightarrow O_P^7(-4) \longrightarrow O_P \longrightarrow O_C \longrightarrow 0 .$$

Un cas particulier de cette courbe est le lieu de contact de deux surfaces de Kummer tangentes le long de leur intersection.

Les courbes situées sur une surface cubique forment un fermé de dimension 31, localement défini par une équation dans H_8^5 ; ce fermé a deux composantes irréductibles formées l'une des courbes canoniques et l'autre des courbes trigonales ou hyperelliptiques.

B) Une courbe canonique non trigonale a la résolution

$$0 \to O_P(-7) \to O_P^2(-6) \oplus O_P^2(-5) \to O_P^3(-4) \oplus O_P(-3) \to O_P \to O_C \to 0$$

Elle est liée par une surface cubique et une surface quartique à la réunion disjointe de deux coniques. La surface cubique n'a que des singularités isolées. En effet, le plongement complet de la courbe est dans $\mathbb{P}^4$, et comme C n'est pas trigonale, tout centre de projection de $\mathbb{P}^4$ appartient à une surface de Del Pezzo de degré 4 contenant la courbe, qui se projette sur la surface cubique.

C) Une courbe trigonale non canonique présente la résolution

$$0 \to O_P(-7) \to O_P^2(-6) \oplus O_P^3(-5) \to O_P(-5) \oplus O_P^3(-4) \oplus O_P(-3) \to O_P \to O_C \to 0$$

Cette courbe a toujours une 5-sécante. En effet, soit L le faisceau de degré 3 et de dimension 2 rendant la courbe trigonale; alors $L^{-1}(1)$ est de degré 5 , donc a une section constituant un groupe de cinq points alignés. Il est amusant de constater que la résolution de cette 5-sécante L apparaît, tordue par -5 , dans la résolution de la courbe. Plus précisement la suite exacte

$$0 \longrightarrow O_L(-5) \longrightarrow O_{CUL} \longrightarrow O_C \longrightarrow 0$$

donne la résolution de O_C comme "mapping cone" du morphisme entre les résolutions de $O_L(-5)$ et O_{CUL} (la courbe $C \cup L$ étant liée a une cubique gauche sur la surface cubique). La surface cubique contenant C a au plus des singularités isolées.

D) Une courbe canonique trigonale présente numériquement la même résolution que la courbe précédente, mais la surface cubique la contenant a une droite double puisque c'est la projection de la surface cubique gauche de $\mathbb{P}^4$ qui contient la courbe canonique trigonale.

On remarque que les courbes B), C) et D) ont la même postulation, les deux dernières présentant le même type numérique. C'est l'exemple de plus bas degré de deux courbes C et C' possédant numériquement la même résolution et telles que $e(C) \neq e(C')$.

E) Une courbe hyperelliptique est une correspondance (2.6) sur une quadrique lisse; elle présente la résolution

$$0 \longrightarrow O_P^3(-8) \longrightarrow O_P^8(-7) \longrightarrow O_P^5(-6) \oplus O_P(-2) \longrightarrow O_P \longrightarrow O_C \longrightarrow 0 .$$

Le schéma de Hilbert $H_{10}^6 = H$ des courbes lisses connexes de degré 10 et genre 6 est normal, connexe, de dimension 40. Les courbes C telles que $e(C) = 1$, i.e. telles que $O_C(1)$ est un module dualisant, forment un fermé irréductible H_s de dimension 38. Les courbes contenues dans une surface quartique forment un fermé H' localement défini dans H par une équation non nulle et contenant H_s. Parmi ces assertions, seule nous semble poser des problèmes la double inclusion stricte $U \underset{\neq}{\supset} H' \underset{\neq}{\supset} H_s$. Pour la démontrer, il suffit de trouver une courbe de H non contenue dans une surface quartique, et de prouver que toute courbe canonique de H est contenue dans une surface quartique.

Considérons d'abord la configuration suivante de quinze droites: On prend cinq points, non quatre à quatre coplanaires, A,B,C,D,E de P . On note (A,BC,DE) la droite passant par A et rencontrant les droites BC et DE . En permutant les cinq points on trouve ainsi 15 droites. Cette configuration contient 15 points singuliers; chaque droite passe par 3 de ces points, et par chacun de ces points passent 3 droites. Les 15 droites ne sont pas contenues dans une surface quartique (celle-ci contiendrait trop de plans). On retire 5 droites deux à deux disjointes, par exemple (A,BC,DE),(B,CD,AE), (C,AD,BE),(D,AB,CE) et (E,AC,BD). On vérifie que les 10 droites restantes ne sont pas contenues dans une surface quartique (car celle-ci contiendrait les 15 droites). Comme le genre arithmétique de cette nouvelle configuration Γ est 6 , la résolution de son cône projetant sera

$$0 \longrightarrow O_P^5(-7) \longrightarrow O_P^{15}(-6) \longrightarrow O_P^{11}(-5) \longrightarrow O_P \ . \qquad (*)$$

Puisque Γ n'a que des points doubles ordinaires et est intersection de surfaces quintiques, il existe une quintique lisse S contenant Γ ; de plus si L est le O_S-module inversible dont Γ est une section, L est engendré par ses sections. Une telle section suffisamment générale est une courbe lisse connexe dont le cône projetant admet une résolution de type numérique (*) , ce qui démontre $H \supsetneqq H'$ ..

Considérons maintenant une courbe canonique de genre 6 dans $\mathbb{P}^3$. C'est la projection d'une courbe arithmétiquement normale de $\mathbb{P}^5$. Pour une projection suffisamment générale de cette courbe dans $\mathbb{P}^2 = \mathrm{Proj}(k[Y_o,Y_1,Y_2])$, le cône $\sum_{n\geq o} H^o(O_C(n))$ est engendré par ses éléments de degré 1 sur $k[Y_o,Y_1,Y_2]$ ([11], p. 56,57). La résolution de ce cône sur $O_{\mathbb{P}^2}$ induit donc une suite exacte

$$0 \longrightarrow O_{\mathbb{P}^2}(-4) \oplus O^3_{\mathbb{P}^2}(-3) \longrightarrow O^3_{\mathbb{P}^2}(-1) \oplus O_{\mathbb{P}^2} \longrightarrow O_C \longrightarrow 0 \ .$$

On en déduit que si C est assez générale dans $\mathbb{P}^3$ la clôture intégrale du cône projetant de C est un $k[X_0,X_1,X_2]$-module engendré par 4 éléments, donc que C est sur une surface quartique. A fortiori toute spécialisation de C et finalement toute courbe canonique de genre 6 dans $\mathbb{P}^3$ est dans une surface quartique.

Nous avons vu que (*) donne le type numérique de la courbe la plus générale de H^6_{10}. Des techniques numériques qu'il serait fastidieux de développer ici montrent que si C est une courbe de H^6_{10} contenue dans une seule quartique et n'admettant ni droite 6-sécante, ni conique 10-sécante, le cône projetant de C admet la résolution

$$0 \to O_P(-8) \oplus O_P(-7) \to O_P^9(-6) \to O_P^7(-5) \oplus O_P(-4) \to O_P \to O_C \to 0.$$

Compte tenu de ce qui a été dit précédemment, pour montrer qu'il existe des courbes présentant ce type avec $e = 0$ ou $e = 1$, il suffit de montrer qu'il en existe avec $e = 1$. Pour ce faire, on considère une courbe canonique de genre 6, arithmétiquement normale, assez générale dans $\mathbb{P}^5$. C'est l'intersection d'une surface de Del Pezzo et d'une hyperquadrique de $\mathbb{P}^5$. On obtient la courbe cherchée en projetant cette courbe dans $\mathbb{P}^3$ à partir d'une droite suffisamment générale de $\mathbb{P}^5$.

Nous n'entreprendrons pas ici la recherche des types numériques des courbes moins générales de H^6_{10}.

Bibliographie.

[1] W. Barth Some properties of stable rank-2 vector bundles on $\mathbb{P}^n$.
Math. Ann. 226, 125-150 (1977).

[2] Bourbaki. Algèbre chap. III (ancienne édition).

[3] C. Castelnuovo. Sui multipli di una serie lineare ...
Rend. circ. Mat. Palermo, t. VII, 1893.

[4] G. Ellingsrud. Sur le schéma de Hilbert des variétés de codimension 2 dans $\mathbb{P}^e$ à cône de Cohen-Macaulay.
Ann. Sc. Ec. Norm. Sup. t. 8, fasc. 4 (1975), p. 423-431.

[5] F.R. Gantmacher. Matrizenrechnung. Vol. 2.
VEB Deutscher Verlag der Wissenschaften.

[6] G. Halphen. Mémoire sur la classification des courbes gauches algébriques.
Oeuvres completes t. III.

[7] J. Harris. Thesis, Harvard University 1977 (preprint).

[8] O.A. Laudal. A generalized tri-secant lemma.
Proc. Tromsø Conference on alg. geo. (July 1977).
To appear (Springer).

[9] Th. Muir. A treatise on the theory of determinants. Dover.

[10] D. Mumford. Curves on algebraic surfaces.
Annals of Math. studies, 59. Princeton.

[11] D. Mumford. Varieties defined by quadratic equations.
Questions on algebraic varieties C.I.M.E, 1970.

[12] C. Peskine, L. Szpiro. Liaison des varietes algebriques.
Inventiones math. 26, 271-302 (1973).

DEFORMATION AND STRATIFICATION OF SECANT STRUCTURE

by

Audun Holme (Bergen, Norway)

Contents

§0. Introduction.

In 1969 I presented a short communication, intended to be of a preliminary nature, at Matematisk Seminar, Oslo [H1]. The objective was to study the variation of the secant schemes in a family of embedded, projective schemes. Some methods associated with this had been employed ad hoc over an affine base in my Ph.D. thesis from 1968, [H2]. The plan was then to develop these ideas further as part of a

---

This work was supported by the Norwegian Research Council for Science and the Humanities. The author wishes to express his gratitude for the hospitality extended at the Massachusetts Institute of Technology, and particularly would like to thank Ursula for the outstanding and very rapid typing of the manuscript.

more comprehensive project, to include for example "projection theorems" of the usual type, but in the relative situation this time. However, the continuation took a different course, and in the papers [H4] - [H7], [H-R], the emphasis is more on embedding-obstruction, characteristic classes and projective invariants. (Of course there are a number of other contributors to this area, say [J], [Lℓ1] - [Lℓ3], [Lak 1] - [Lak 3], [P.-S.], [R1] - [R4]. A survey of this and other work may be found in [K].) Thus the deformation of secant structure was almost completely left out, and only touched in [H3] for infinitesimal deformations.

Recently various people have expressed interest in this material, as well as in related concepts and phenomena. Also there have appeared several articles which have a direct bearing on these questions, of which I will particularly single out [R1], [F-M] and [Lau].

The present article represents the (long overdue) definitive version of [H1]. Needless to say, it has become rather different from the form it would have had around 1970. Perhaps the main point to mention is a realization that the "secant structure" of an embedded scheme should include two sets of embedded Segre-classes and numerical invariants associated with them.

Moreover, some of the constructions given in [H1] have appeared in full detail elsewhere, and thus will not be repeated here. In particular this applies to the proofs of (1.1) and Proposition 2.1.

For simplicity we assume that all schemes are quasi-projective over an algebraically closed field. The generalization to noetherian schemes should present no essential difficulties, and in fact some of the proofs have been formulated so that they apply to the general case.

§1. The classical case.

Let $i : X \hookrightarrow \mathbb{P}^N_k = \mathbb{P}^N$ be an embedded, projective scheme over the algebraically closed field k. A secant to X in $\mathbb{P}^N$ is a line meeting X in more than one point, counted with "multiplicity."

This notion can be made precise in at least two ways. In fact, one obtains a commutative diagram

(1.1)
$$\begin{array}{ccccc} T & \hookrightarrow & B & \xrightarrow{\lambda} & \mathbb{P}(\Omega^1_{\mathbb{P}^N}) = T \\ \downarrow \phi & & \downarrow \pi & & \downarrow \phi \\ \mathbb{P}^N & \hookrightarrow & \mathbb{P}^N \times \mathbb{P}^N & \xrightarrow{pr_2} & \mathbb{P}^N \end{array}$$

where π is the blowing-up of the diagonal, ϕ the canonical projection and λ a certain $\mathbb{P}^1$-bundle. (See [H1] or [H6] for one proof, [Lak 1] for another.)

One now puts

$$\lambda(\widetilde{X \times X}) = Sb(X)$$

where $\widetilde{X \times X}$ is the strict transform of $X \times X$ under π, and one denotes the morphism induced from f by

$$s_X : Sb(X) \to X .$$

We call s_X the <u>secant bundle</u> of the (embedded) scheme X. Further,

(1.2)
$$Sec(X) = pr_1(\pi(\lambda^{-1}(Sb(X))))$$

is easily seen to be the closure of the union of all lines with at least two distinct points in common with X. $Sec(X)$ is thus referred to as the (embedded) secant scheme of X.

Another approach, which makes obvious the natural stratification of the secant structure, is via grassmanians. In fact, more generally one studies the system of q-dimensional linear subspaces of $\mathbb{P}^N$ with r or more points in common with X: Let $G = G(q,N)$ be the grassmanian which parametrizes the q-dimensional linear subspaces, $\Gamma = \Gamma(r,q,N) \subset (\mathbb{P}^N)^r \times G$ the incidence correspondence and $V \subset X^r$ the complement of the union of all (multi) diagonals. Define

$$(1.3) \qquad G(X,q)_r = pr_{r+1}(V \times G \cap \Gamma)$$

where (as always) intersection and image are scheme-theoretic. Finally let $p_1 : \Gamma(1,q,N) \to \mathbb{P}^N$ and $p_2 : \Gamma(1,q,N) \to G$ be induced by the projections. Then define

$$(1.4) \qquad Sec(X,q)_r = p_1(p_2^{-1}(G(X,q)_r))$$

which is the closure of the union of all $\mathbb{P}^q$'s in $\mathbb{P}^N$ meeting X in r or more points. Clearly

$$(1.5) \qquad G(X,q)_r \supseteq G(X,q)_{r+1}$$

which in particular yields the above-mentioned stratification.

For more details and a further study of this, see [Lℓ 2]. For instance, Lluis proves for $r \leq q + 1$

$$(1.6) \qquad \dim Sec(X,q)_r \leq (q - r + 1)(N - q) + q + r \dim(X).$$

As for the relation between the two approaches, we have that as sets

(1.7) $$\mathrm{Sec}(X,1)_2 = \mathrm{Sec}(X),$$

with equality as schemes if X is reduced and irreducible.* In fact, set theoretically this is obvious. For the rest, recall the commutative diagram

(1.8)

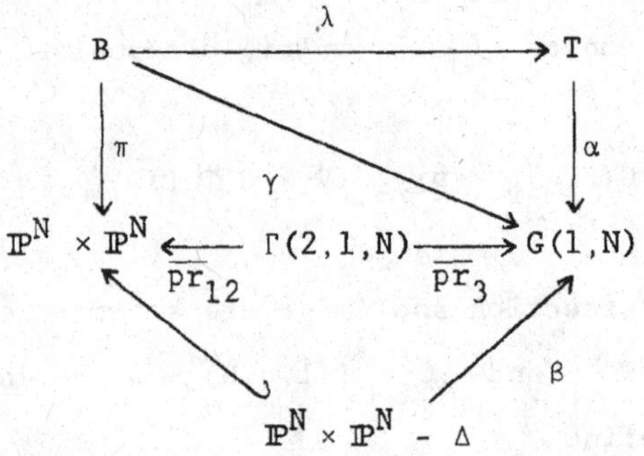

see [K] or [Lak 1] for a nice explanation. β maps a point (x,y) off the diagonal Δ to the point in $G(1,N)$ assigned the line through x and y. $\overline{pr}_{12}$ and $\overline{pr}_3$ are the morphisms induced by the projections. One immediately gets that as schemes, $G(X,1)_2 = \beta(X \times X - \Delta) = \gamma(\widetilde{X \times X})$, so

(1.9) $$G(X,1)_2 = \alpha(\mathrm{Sb}(X)).$$

We next use the diagram

---

* Actually, I expect this to be true in general, but a proof would require a more refined analysis.

$$
(1.10)\qquad
\begin{array}{ccccc}
B & & \xrightarrow{\ \lambda\ } & & T \\
\big\downarrow{\scriptstyle \pi} & & & & \\
\mathbb{P}^N \times \mathbb{P}^N & & & & \big\downarrow{\scriptstyle \alpha} \\
\big\downarrow{\scriptstyle pr_1} & & & & \\
\mathbb{P}^N & \xleftarrow[p_1]{} & \Gamma(1,1,N) & \xrightarrow[p_2]{} & G(1,N)
\end{array}
$$

where $\alpha(z) = x$ implies that

$$
(1.11)\qquad pr_1(\pi(\lambda^{-1}(z))), \quad p_1(p_2^{-1}(x))
$$

are the same line in $\mathbb{P}^N$. Now (1.10) and (1.9), together with (1.2) and (1.4) gives the scheme-theoretic part of the claim.

§2. The relative situation.

Now we carry out the construction in §1 for the relative situation for an "embedded projective morphism" $f : X \to Y$, i.e., for a factorization of f as a closed embedding i followed by the canonical morphism p:

$$
\begin{array}{ccc}
X & \xhookrightarrow{\ i\ } & \mathbb{P}^N_Y \\
& {\scriptstyle f}\searrow \quad \swarrow{\scriptstyle p} & \\
& Y &
\end{array}
$$

To do this, we need a global version of the diagram (1.1). In fact, we have the

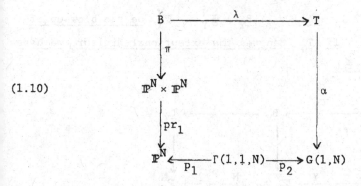

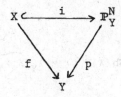

PROPOSITION 2.1. *Let* $\pi_Y : \underline{B}_Y \to \mathbb{P}_Y^N \times_Y \mathbb{P}_Y^N$ *be the blow-up of the* (Y-) *diagonal. If* $\underline{T}_Y$ *denotes the exceptional divisor, we have a commutative diagram*

$$(2.1.1) \qquad \begin{array}{ccccc} \underline{T}_Y & \xhookrightarrow{\varepsilon_Y} & \underline{B}_Y & \xrightarrow{\lambda_Y} & \underline{T}_Y \\ \downarrow{\scriptstyle \phi_Y} & & \downarrow{\scriptstyle \pi_Y} & & \downarrow{\scriptstyle \phi_Y} \\ \mathbb{P}_Y^N & \xhookrightarrow{\delta_Y} & \mathbb{P}_Y^N \times_Y \mathbb{P}_Y^N & \xrightarrow{pr_2} & \mathbb{P}_Y^N \end{array}$$

Here λ_Y *is locally a product with* $\mathbb{P}_Y^1$, δ_Y *the diagonal embedding,* ε_Y *the canonical closed embedding and* ϕ_Y *the morphism induced by* π_Y.

Moreover, the diagram is compatible with any base extension of Y

Proof: As $\mathbb{P}_Y^N$ is smooth over Y, $\Omega^1_{\mathbb{P}_Y^N/Y}$ is locally free. Hence $\underline{T}_Y = \mathbb{P}(\Omega^1_{\mathbb{P}_Y^N/Y})$, ϕ_Y being the canonical projection. Similarly this blowing-up commutes with base extension of Y (see for instance the proof of Proposition 3.5).

Thus it remains to construct the $\mathbb{P}_Y^1$-bundle λ_Y such that the diagram commutes. This is not hard, indeed one observes that the proof of (1.1) given in [H1] and [H6] applies.

Letting $\widetilde{X \times_Y X}$ denote the strict transform of $X \times_Y X$, we define

$$Sb(X/Y,i) = \lambda_Y(\widetilde{X \times_Y X}) .$$

ϕ_Y induces a morphism

$$s_{X/Y} : Sb(X/Y,i) \to X .$$

We delete i when no confusion is possible. This morphism is referred to as the secant-bundle of the embedded, projective Y-scheme X. If $Y = \mathrm{Spec}(k)$, one gets the construction from §1. Finally one notes that of course $s_{X/Y}$ is by no means locally trivial in general, rendering the term "bundle" somewhat misleading. The same situation exists in §1, the justification for the terminology being that the secant bundle parametrizes all secant-directions at a given point of X, in the same way as the cotangent bundle does tangent directions.

Similarly one globalizes the other concepts in §1. Let $G_Y = G \times Y$, and $\Gamma_Y = \Gamma \times Y$. V_Y denotes the complement of the union of all (multi) diagonals in X^r. In particular we have the two diagrams

(2.2)
$$\begin{array}{ccccc} & & \Gamma_Y(r,q,N) & & \\ & \swarrow^{p_1} & & \searrow^{p_2} & \\ (\mathbb{P}^N_Y)^r & & & & G_Y(q,N) \end{array} \qquad \begin{array}{ccccc} & & \Gamma_Y(1,q,N) & & \\ & \swarrow^{p_1} & & \searrow^{p_2} & \\ \mathbb{P}^N_Y & & & & G_Y(q,N) \end{array}$$

where the morphisms are induced by the projections. Define

$$G(X/Y,q)_r = p_2(p_1^{-1}(V_Y)) \tag{2.3}$$

and

$$\mathrm{Sec}(X/Y,q)_r = p_1(p_2^{-1}(G(X/Y,q)_r)). \tag{2.4}$$

(Again, image by a morphism is always the (closed) scheme-theoretic one.)

Lluis' estimate (1.6) can be generalized to this situation, in fact assuming that $r \leq q + 1$ we have

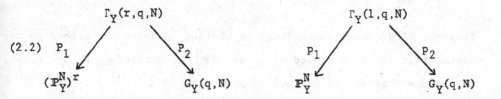

$$\text{(2.5)} \qquad \dim \mathrm{Sec}(X/Y,q)_r \leq (q-r+1)(N-q) + q + r \dim X - (r-1)\dim Y$$

The proof is that of Lluis, with some obvious modifications. Namely, let $x \in (\mathbb{P}^N_Y)^r$ and $y \in Y$ be its canonical image. Then

$$\text{(2.6)} \qquad \dim P_1^{-1}(x) = \dim P_{1,K}^{-1}(x) \leq (q-r+1)(N-q)$$

where the middle entity refers to the fibers over y, $K = k(y)$:

$$\text{(2.7)} \qquad \begin{array}{ccc} & \Gamma_K(r,q,N) & \\ {}^{P_{1,K}}\swarrow & & \searrow^{P_{2,K}} \\ (\mathbb{P}^N_K)^r & & G_K(q,N) \end{array}$$

Thus the first equality in (2.6) is obvious, and for the second we may assume that x is a closed point of $(\mathbb{P}^N_K)^r$. Replacing K by $\bar{K}$, we may assume that x is a K-point, and then

$$P_{1,K}^{-1}(x) \cong G_K(q - (r'+1), N - (r'+1))$$

where r' is the dimension of the linear subspace in $\mathbb{P}^N_K$ spanned by the r projections of x. Thus

$$\dim P_{k,K}^{-1}(x) = (q-r')(N-q) \leq (q-r+1)(N-q),$$

proving (2.6).

Similarly one finds $\dim p_2^{-1}(z) = q$. Finally noting that $\dim V_Y = r \dim X - (r-1)\dim Y$, (2.5) follows.

(1.7) also holds in the relative situation: As sets,

$$\text{(2.8)} \qquad \mathrm{Sec}(X/Y,1)_2 = \mathrm{Sec}(X/Y).$$

To see this, one notes that (1.8) holds over Y, say by extending the base to Y. One then obtains (1.9) in the relative case, and completes the proof by (1.10) extended to Y, where (1.11) takes place in $\mathbb{P}^N_K$, $K = k(y)$.

§3. Deforming the secant bundle.

Here we limit ourselves to $Sb(X/Y)$ and $Sec(X/Y)$. To study their behaviour under deformation, we need some general information first.

In fact, neither blowing-up nor scheme theoretic image are compatible with base extension in general. We need formulas expressing the difference.

Let $\pi : \tilde{X} \to X$ be the blow-up of the S-scheme X along a subscheme Y, and $\phi : T \to S$ be a base extension. One obtains a canonical morphism

$$h : \widetilde{(X_T)} \to \tilde{X}_T$$

where $\widetilde{(X_T)}$ is the blow-up along Y_T. We determine how far h is from being an isomorphism. So let I be the Ideal defining Y, then

$$\tilde{X}_T = \mathrm{Proj}(\mathrm{Pow}_{O_X}(I) \otimes_{O_S} O_T).$$

Y_T is defined by $I' = \mathrm{im}(I \otimes_{O_S} O_T \to O_X \otimes_{O_S} O_T = O_{X_T})$, thus

$$\widetilde{(X_T)} = \mathrm{Proj}(\mathrm{Pow}_{O_{X_T}}(I')).$$

PROPOSITION 3.1. h <u>is a closed embedding</u>, <u>and identifies</u> $\widetilde{(X_T)}$ <u>with the closed subscheme of</u> $\tilde{X}_T$ <u>given by an Ideal</u> $\tilde{\mathfrak{a}}$, <u>where</u> $\mathfrak{a}$ <u>is a</u>

<u>graded Ideal in</u> $Pow_{O_X}(I) \otimes_{O_S} O_T$. <u>Provided the infinite sums have meaning, the following formula holds in</u> K.T <u>for all</u> $n \geq 1$:

$$[\alpha_n] = \sum_{i=1}^{\infty} (-1)^i [Tor_i^{O_S}(O_X,O_T)] - \sum_{i=1}^{\infty} (-1)^i [Tor_i^{O_S}(O_X/I^n,O_T]$$
$$- \sum_{i=1}^{\infty} (-1)^i [Tor_i^{O_S}(I^n,O_T)]$$

<u>Proof</u>: Since $I'^n = im(I^n \otimes_{O_S} O_T \to O_{X_T})$ the morphism h is defined by the homomorphism

$$\phi : Pow_{O_X}(I) \otimes_{O_S} O_T \to Pow_{O_{X_T}}(I')$$

which is induced by

$$\phi_n : I^n \otimes_{O_S} O_T \to I'^n.$$

Thus h is a closed embedding. Moreover,

$$\alpha_n = \ker(\phi_n).$$

Now the short exact sequence

$$0 \longrightarrow I^n \longrightarrow O_X \longrightarrow O_X/I^n = O_X^{(n)} \longrightarrow 0$$

yields the long exact Tor-sequence

$$I^n \otimes_{O_S} O_T \longrightarrow O_X \otimes_{O_S} O_T \longrightarrow O_X^{(n)} \otimes_{O_S} O_T \longrightarrow 0$$
$$Tor_1^{O_S}(I^n,O_T) \longrightarrow Tor_1^{O_S}(O_X,O_T) \longrightarrow Tor_1^{O_S}(O_X^{(n)},O_T)$$
$$\cdots \longrightarrow Tor_2^{O_S}(O_X^{(n)},O_T)$$

for all $n \geq 1$. Thus one obtains the exact sequence

$$\operatorname{Tor}_1^{O_S}(I^n, O_T) \longrightarrow \operatorname{Tor}_1^{O_S}(O_X, O_T) \longrightarrow \operatorname{Tor}_1^{O_S}(O_X^{(n)}, O_T) \longrightarrow \mathfrak{a}_n \longrightarrow 0$$

$$\cdots \longrightarrow \operatorname{Tor}_2^{O_S}(O_X^{(n)}, O_T)$$

from which the claim follows.

COROLLARY 3.1.1. <u>If</u> T <u>is flat over</u> S, <u>then</u> $\widetilde{(X_T)} = \tilde{X}_T$. <u>If</u> X <u>if flat over</u> S, <u>then</u>

$$\mathfrak{a}_n \cong \operatorname{Tor}_1^{O_S}(O_X^{(n)}, O_T).$$

The other basic proposition needed concerns the behaviour of scheme-theoretic image under base extension. Let $f : X \to Y$ be a morphism of S-schemes, and $\phi : T \to S$ be a base extension. We have the diagram

$$\begin{array}{ccc} X' = X_T & \xrightarrow{\quad g' \quad} & X \\ \Big\downarrow{\scriptstyle f_T = f'} & & \Big\downarrow{\scriptstyle f} \\ Y' = Y_T & \xrightarrow[\quad g \quad]{} & Y \end{array}$$

and for each coherent O_X-Module F the canonical homomorphism

$$\tau_{F,\phi} : g^* f_* F \to f'_* g'^* F$$

associated with the base change ϕ.

If f is an affine morphism, or if ϕ is flat, then $\tau_{F,\phi}$

is an isomorphism (EGA III, 1.4.14 and 1.4.15). But in general it is not, indeed the Module $\boldsymbol{\tau}_\phi = \ker \tau_{O_X,\phi}$ does to some extent measure the change induced by the base extension. In fact, letting

$$\boldsymbol{\tau}_\phi(f) = \boldsymbol{\tau}_\phi \cap \operatorname{im}(O_{f(X)} \otimes_{O_Y} O_{Y'} \to f_* O_X \otimes_{O_Y} O_{Y'})$$

and writing $f(X)_T = f(X)'$, we have the

PROPOSITION 3.2. <u>There is a canonical closed embedding</u> $\ell : f'(X') \hookrightarrow f(X)'$, <u>inducing an isomorphism of the reduced subschemes if</u> f <u>is proper</u>. <u>Further, if</u> I <u>defines</u> $f(X)$ <u>then</u>

(3.2.1)
$$[f(X)'] = [Y'] - [I \otimes_{O_Y} O_{Y'}] + [\mathrm{Tor}_1^{O_Y}(O_{f(X)}, O_{Y'})]$$

<u>and if the infinite sums have meaning</u>,

(3.2.2)
$$[f(X)'] - [f'(X')] = [\boldsymbol{\tau}_\phi(f)] + \sum_{i=1}^{\infty} (-1)^i [\mathrm{Tor}_i^{O_Y}(f_* O_X, O_{Y'})]$$
$$- \sum_{i=1}^{\infty} (-1)^i [\mathrm{Tor}_i^{O_Y}(f_* O_X / O_{f(X)}, O_{Y'})] - \sum_{i=1}^{\infty} (-1)^i [\mathrm{Tor}_i^{O_Y}(O_{f(X)}, O_{Y'})]$$

<u>Proof</u>: We first note the commutative diagram

(3.2.3)
$$\begin{array}{ccc} O_{Y'} & \xrightarrow{\quad\Phi\quad} & f'_*(O_{X'}) \\ \downarrow{\scriptstyle\theta''} & \searrow^{\theta'} & \uparrow{\scriptstyle\tau} \\ O_{f(X)} \otimes_{O_Y} O_{Y'} = O_{f(X)'} & \xrightarrow{\quad\psi\quad} & f_* O_X \otimes_{O_Y} O_{Y'} \end{array}$$

where $\theta' = \theta \otimes_{O_Y} O_{Y'}$, $\theta : O_Y \to f_* O_X$ being associated to f. Further, $\theta'' = \bar{\theta} \otimes_{O_Y} O_{Y'}$ where $\bar{\theta} : O_Y \to \operatorname{im} \theta = O_{f(X)}$ is induced by θ. Finally ψ is $\otimes_{O_Y} O_{Y'}$ applied to the canonical injection, while Φ is associated to f'.

The subschemes $f(X)'$ and $f'(X')$ are given by the Ideals $\ker(\theta'') = \operatorname{im}(I \otimes_{O_Y} O_{Y'} \to O_{Y'})$ and $\ker(\Phi)$, respectively. As $\ker(\Phi) \supseteq \ker(\theta'')$, we get ℓ.

If f is proper the induced morphism $\bar{f} : X \to f(X)$ is surjective This being preserved by base extension (EGA I, 3.5.2) we conclude that $f(X)'$ has the same underlying set as $f'(X')$.

The short exact sequence $0 \to I \to O_Y \to O_{f(X)} \to 0$ yields a Tor-sequence which amounts to

$$0 \to \operatorname{Tor}_1^{O_Y}(O_{f(X)}, O_{Y'}) \to I \otimes_{O_Y} O_{Y'} \to O_{Y'} \to O_{f(X)} \otimes_{O_Y} O_{Y'} \to 0$$

as well as the isomorphisms

$$\operatorname{Tor}_i^{O_Y}(I, O_{Y'}) \cong \operatorname{Tor}_{i+1}^{O_Y}(O_{f(X)}, O_{Y'}), \quad i \geq 1.$$

Thus (3.2.1) follows.

By (3.2.3) $\ker(\Phi) = \theta'^{-1}(\ker(\tau))$, so one obtains

$$\begin{array}{ccccccc} 0 & \longrightarrow & \ker(\theta') & \longrightarrow & \ker(\Phi) & \longrightarrow & \ker(\tau) \\ & & \| & & \downarrow & & \downarrow \\ 0 & \longrightarrow & \ker(\theta') & \longrightarrow & O_{Y'} & \xrightarrow{\theta'} & f_*O_X \otimes_{O_Y} O_{Y'} \end{array}$$

Since $\operatorname{im} \theta' = \operatorname{im}(\psi)$, this yields

$$(3.2.4) \qquad 0 \to \ker(\theta') \to \ker(\Phi) \to \boldsymbol{\tau}_\phi(f) \to 0$$

As θ'' is surjective, a similar argument proves

$$(3.2.5) \qquad 0 \to \ker(\theta'') \to \ker(\theta') \to \ker(\psi) \to 0$$

Hence

$$[f(X)'] - [f'(X')] =$$

$$[O_{Y'}] - [\ker(\theta'')] - ([O_{Y'}] - [\ker(\Phi)]) =$$

$$[\ker(\Phi)] - [\ker(\theta'')] =$$

$$[\ker(\theta')] + [\tau_\phi(f)] - ([\ker(\theta')] - [\ker(\psi)]) =$$

$$[\tau_\phi(f)] + [\ker(\psi)]$$

To compute $[\ker(\psi)]$ and thus complete the proof, one now uses the Tor-sequence obtained from

$$0 \to O_{f(X)} \to f_*O_X \to f_*O_X/O_{f(X)} \to 0 ,$$

namely

$$O_{f(X)} \otimes_{O_Y} O_{Y'} \longrightarrow f_*O_X \otimes_{O_Y} O_{Y'} \longrightarrow f_*O_X/O_{f(X)} \otimes_{O_Y} O_{Y'} \longrightarrow 0$$

$$\mathrm{Tor}_1^{O_Y}(O_{f(X)},O_{Y'}) \longrightarrow \mathrm{Tor}_1^{O_Y}(f_*O_X,O_{Y'}) \longrightarrow \mathrm{Tor}_1^{O_Y}(f_*O_X/O_{f(X)},O_{Y'})$$

$$\cdots$$

which yields the long exact sequence

$$\mathrm{Tor}_1^{O_Y}(O_{f(X)},O_{Y'}) \to \mathrm{Tor}_1^{O_Y}(f_*O_X,O_{Y'}) \to \mathrm{Tor}_1^{O_Y}(f_*O_X/O_{f(X)},O_{Y'}) \to \ker(\psi) \to 0$$

$$\cdots$$

from which the claim follows.

<u>Remark 3.2.6</u>. In particular $f'(X') = f(X)'$ if ϕ is flat,

and

$$[f'(X')] - [f(X)'] = [\tau_\phi(f)] + [\mathrm{Tor}_1^{O_Y}(f_*O_X/O_{f(X)}, O_{Y'})]$$

is f is flat. $[\tau_\phi(f)] = 0$ whenever f is affine.

The following formula is strictly speaking not needed in the sequel. But it completes the picture, and thus should be included.

Let $f : Z \to Y$ be a morphism associated with $\theta : O_Y \to f_*O_Z$, and X be the subscheme defined by the Ideal I. Then the following holds in K.Y:

$$(3.3) \qquad [f(X)] - f_![X] = [f(Z)] - f_![Z] + \sum_{i=1}^{\infty} (-1)^i [R^i f_* I]$$

$$+ [f_*I/\mathrm{im}\,\theta \cap f_*I]$$

In fact, the short exact sequence $0 \to I \to O_Z \to O_X \to 0$ yields

$$0 \to f_*I \to f_*O_Z \to f_*O_X \to R^1 f_*I \to R^1 f_*O_Z \to R^1 f_*O_X \to \cdots$$

which we split in two:

$$(3.3.1) \qquad 0 \to f_*I \to f_*O_Z \to K \to 0$$

$$(3.3.2) \qquad 0 \to K \to f_*O_X \to R^1 f_*I \to R^1 f_*O_Z \to R^1 f_*O_X \to \cdots$$

(3.3.1) implies the following diagram of exact sequences:

$$
\begin{array}{ccccccccc}
 & & 0 & & 0 & & 0 & & \\
 & & \uparrow & & \uparrow & & \uparrow & & \\
0 & \longrightarrow & \operatorname{coker}(\bar{\theta}) & \longrightarrow & \operatorname{coker}(\theta) & \longrightarrow & K/O_{f(X)} & \longrightarrow & 0 \\
 & & \uparrow & & \uparrow & & \uparrow & & \\
0 & \longrightarrow & f_*I & \longrightarrow & f_*O_Z & \longrightarrow & K & \longrightarrow & 0 \\
 & & \uparrow{\scriptstyle \bar{\theta}} & & \uparrow{\scriptstyle \theta} & & \uparrow & & \\
0 & \longrightarrow & \theta^{-1}(f_*I) & \longrightarrow & O_Y & \longrightarrow & O_{f(X)} & \longrightarrow & 0 \\
 & & \uparrow & & \uparrow & & \uparrow & & \\
0 & \longrightarrow & \ker(\bar{\theta}) & \longrightarrow & \ker(\theta) & \longrightarrow & 0 & & \\
 & & \uparrow & & \uparrow & & & & \\
 & & 0 & & 0 & & & &
\end{array}
\tag{3.3.3}
$$

Hence

$$[K] = [f(X)] + [\operatorname{coker}(\theta)] - [\operatorname{coker}(\bar{\theta})]$$

which by the exact sequence

$$0 \to O_{f(Z)} \to f_*O_Z \to \operatorname{coker}(\theta) \to 0$$

yields

$$[K] = [f(X)] + [f_*O_Z] - [f(Z)] - [\operatorname{coker}(\bar{\theta})] .$$

By (3.3.2) we now get

$$[K] = f_![X] + \sum_{i=1}^{\infty} (-1)^i [R^i f_* I] - \sum_{i=1}^{\infty} (-1)^i [R^i f_* O_Z]$$

and thus

$$[f(X)] - f_![X] = [f(Z)] - [f_!(Z)] + \sum_{i=1}^{\infty} (-1)^i [R^i f_* I] + [\operatorname{coker}(\bar{\theta})]$$

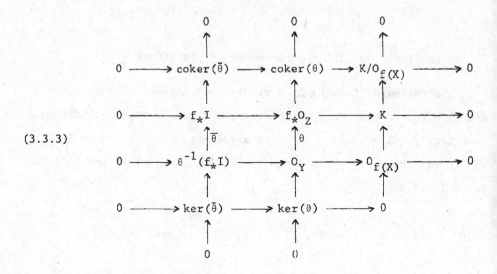

Since coker $\bar{\theta} \cong f_*I/\text{im}\,\theta \cap f_*I$, (3.3) is proven.

In two important special cases one obtains more explicitly the

PROPOSITION 3.4. <u>If</u> $f : Z \to Y$ <u>is either a</u> $\mathbb{P}^r$<u>-bundle, or the blow up of some regularly embedded closed subscheme of</u> Y, <u>then</u>

$$[f(X)] - f_![X] = \sum_{i=1}^{\infty} (-1)^i [R^i f_* I]$$

<u>Proof</u>: In both cases one has

$$R^i f_* O_Z = \begin{cases} O_Y & \text{for } i = 0 \\ 0 & \text{for } i > 0 \end{cases}$$

see EGA III, Proposition 2.1.12 and [M], 14.3. Thus $f_![Z] = [Y]$. Since θ is an isomorphism here, we also have $O_{f(Z)} = O_Y$, so that $[f(Z)] = [Y]$. Moreover, $f_*I/\text{im}\,\theta \cap f_*I = 0$, and the claim is clear.

We return to the secant-bundle of a projective morphism $f : X \to Y$. Let $\phi : Y' \to Y$ be any base extension, and

$$h : \overbrace{X_{Y'} \times_{Y'} X_{Y'}} \hookrightarrow (\overbrace{X \times_Y X})_{Y'}$$

be the closed embedding associated with the base change. This yields the closed embedding

$$\lambda_{Y'}(\overbrace{X_{Y'} \times_{Y'} X_{Y'}}) \hookrightarrow \lambda_{Y'}((\overbrace{X \times_Y X})_{Y'})$$

Further one has the closed embedding

$$\lambda_{Y'}((\overbrace{X \times_Y X})_{Y'}) \hookrightarrow \lambda_Y(\overbrace{X \times_Y X})_{Y'}$$

and the composition yields the <u>base change morphism</u> for the secant bundle, namely the closed embedding

$$b_\phi : Sb(X_{Y'}/Y') \hookrightarrow Sb(X/Y)_{Y'} .$$

A complete description of how the secant bundle behaves under base change would amount to a computation of

$$[Sb(X/Y)_{Y'}] - [Sb(X_{Y'}/Y')] \in K.\underline{T}_{Y'} .$$

While such a computation is now possible in principle, essentially by means of Propositions 3.1 and 3.2, the formula seems to become too involved to be of any practical use. We will, however, return to this question in detail at a later occasion. For the time being we note only that if ϕ is flat, then b_ϕ is an isomorphism, which is immediate from the two propositions referred to above.

On the other hand, one has the

PROPOSITION 3.5. *If the morphism* $f : X \to Y$ *is smooth and* $\phi : Y' \to Y$ *is a base extension with* Y' *reduced, then*

$$Sb(X_{Y'}/Y') = (Sb(X/Y)_{Y'})_{red}$$

Proof: $X_{Y'} \to Y'$ is smooth, thus so is $X_{Y'} \times_{Y'} X_{Y'} \to X_{Y'}$, thus so is $X_{Y'} \times_{Y'} X_{Y'} \to Y'$, EGA IV, 17.3.3. Hence $X_{Y'} \times_{Y'} X_{Y'}$ is reduced, *loc.cit* 17.5.7. It follows that $\widetilde{X_{Y'} \times_{Y'} X_{Y'}}$ is reduced

Moreover, we note that $(\widetilde{X \times_Y X})_{Y'} = \widetilde{X_{Y'} \times_{Y'} X_{Y'}}$ in this case: By flattness (*loc. cit.* 17.5.1) our Corollary 3.1.1 yields

$$\mathfrak{A}_n \cong Tor_1^{O_Y}(P^{n-1}_{X/Y}, O_{Y'})$$

where $P^{n-1}_{X/Y} = O_{X \times_Y X}/I^n$ is the Module of principal parts. This bein locally free, loc. cit. 16.10.1 and 17.12.4, the claim follows.

Thus by EGA I, 9.5.9, the subscheme of $\underline{T}_{Y'}$,

$$\lambda_{Y'}(\widetilde{X \times_Y X_{Y'}}) = \lambda_{Y'}(\widetilde{X_{Y'} \times_{Y'} X_{Y'}}) = \mathrm{Sb}(X_{Y'}/Y')$$

is reduced. Hence the claim of the proposition follows by the first part of Proposition 3.2.

Recall that a geometric point of Y is a morphism τ: $\mathrm{Spec}(K) \to Y$ where K is a field. The corresponding fiber is

$$X_\tau = f^{-1}(\tau) = X \times_Y \mathrm{Spec}(K)$$

while the scheme $f^{-1}(\tau)_{red}$ is referred to as the reduced fiber of f over τ. The following is an immediate corollary of the last proposition.

THEOREM 3.6. *Let $f : X \to Y$ be a smooth, projective embedded morphism. Then for all geometric points τ of Y, $\mathrm{Sb}(X_\tau)$ is equal to the reduced geometric fiber of $\mathrm{Sb}(X/Y)$ over τ.*

Remark. It would be interesting to know precisely when one can delete the work "reduced" in the statement above, a similar question applies to Proposition 3.5.

Having determined how the secant bundle varies in a (smooth) family, we finally note that a similar result holds for the secant scheme as well. Let

$$\mathrm{Sec}(X/Y) = pr_1 \pi_Y \lambda_Y^{-1} \mathrm{Sb}(X/Y) \subseteq \mathbb{P}_Y^N .$$

With notations as before, the following holds:

THEOREM 3.7. For all geometric points τ of Y, $\mathrm{Sec}(X_\tau)$ is equal to the reduced fiber of $\mathrm{Sec}(X/Y)$ over τ.

The proof is analogous to that of Theorem 3.6, in fact one proves that in the situation of Proposition 3.5,

$$\mathrm{Sec}(X_{Y'}/Y') = (\mathrm{Sec}(X/Y)_{Y'})_{\mathrm{red}} \tag{3.8}$$

This is shown in the same manner as Proposition 3.5.

Remark. It would be interesting to know if results of this type hold for the generalizations of Sb( ) and Sec( ) studied in §1, 2. This seems to require a more refined analysis of the situation than what is presently possible.

§4. Enumerative data and their variation.

Following [J], [F-M] and [K2] we define two types of Segre-classes for the projective morphism $f : X \to Y$. Recall the diagram of canonical morphisms

$$\begin{array}{ccc} T(X/Y) & \hookrightarrow & P(X/Y) = \mathbb{P}(\Omega^1_{X/Y}) \\ & {\scriptstyle\tau}\searrow \quad \swarrow{\scriptstyle\pi} & \\ & X & \end{array} \tag{4.1}$$

where $T(X/Y)$ is the exceptional divisor of $\widetilde{X \times_Y X}$. Then with the notations of [F-M],

$$s(X/Y) = \tau_*(c(\mathcal{O}_{T(X/Y)}(-1))^{-1} \wedge [T(X/Y)])$$

$$\sigma(X/Y) = \pi_*(c(\mathcal{O}_{P(X/Y)}(-1))^{-1} \wedge [P(X/Y)])$$

are two (total) Segre-classes of X/Y in the abelian group $A.X$. In general they are different, but if f is differentially smooth (EGA IV, 16.10.1), in particular if it is smooth (EGA IV, 17.12.4), then they coincide. $A.X$ is graded by dimension as well as by co-dimension, the corresponding homogeneous parts are denoted by s_i and s^i, σ_i and σ^i, respectively.

When X is a projective scheme over a field k, then certain interesting numerical data concerning X are expressed in terms of the above invariants. In fact, let $i : X \hookrightarrow \mathbb{P}^N_k$ be a projective embedding over k and $p : X \to X'$ be induced by a generic projection from $\mathbb{P}^N_k$ to $\mathbb{P}^m_k$, X' being the scheme theoretic image. Furthermore, write

$$i_* s_j(X) = p_j(X) t^{N-j}$$

where $t = [H] \in A.\mathbb{P}^N_k$, H being a hyperplane. k is deleted in s, σ when $Y = \mathrm{Spec}(k)$. Similarly,

$$i_* \sigma_j(X) = q_j(X) t^{N-j}$$

Thus p_j, q_j are the degrees of s_j, q_j with respect to the embedding i. Then the number of multiple points as well as the number of pinch points of p, counted with certain natural multiplicities*, are given in terms of these invariants as follows:

$$\#\text{ multiple points} = \begin{cases} \deg(X)^2 - \sum_{j=0}^{m-n} \binom{m+1}{m-n-j} p_j(X) & \text{for } m \leq 2n \\ 0 & \text{otherwise} \end{cases}$$

---

* Actually the degree of a cycle of multiple, respectively pinch, points.

$$\# \text{ pinch points} = \begin{cases} \displaystyle\sum_{j=0}^{m+n-\rho} \binom{m+1}{j+\rho-n+1} q_j(X) & \text{for } \rho - n \leq m \leq \rho \\ 0 & \text{for } m > \rho \end{cases}$$

Here $n = \dim(X)$, $\rho = \dim P(X)$. For proofs, see for instance [H-R] Theorem 3.3. One may also consult [H6], [H7], [J], [K], [Lak 1]-[Lak

It is of interest to determine how these numbers vary in a family of embedded projective schemes: Let $f : X \to Y$ be an embedded projective morphism

$$(4.2) \qquad \begin{array}{ccc} X & \overset{i}{\hookrightarrow} & \mathbb{P}_Y^N \\ & {\scriptstyle f}\searrow \quad \swarrow{\scriptstyle p} & \\ & Y & \end{array}$$

For all geometric points $\tau : \mathrm{Spec}(K) \to Y$ we obtain a closed embeddin $i_\tau : X_\tau \hookrightarrow \mathbb{P}_K^N$. Denoting the two numbers above by $\mathrm{mult}_\tau(m)$, $\mathrm{pinch}_\tau(m)$ respectively, one would like to determine how they vary by explicit formulas. For this we may assume that Y is connected, which we do from now on.

At present no complete solution to the above seems to be known. However, if f is smooth, then it is shown in [R1] that the two numbers above are constant as τ varies (cf. Theorem 1 of [R1]).

Here we present a different approach, depending on generalization of the degrees $p_j(X)$ and $q_j(X)$ to the relative situation. We study how these "degrees" vary under base change, this is closely related to the same question for $\mathrm{Sb}(X/Y)$.

We first list a lemma on the structure of $A.\mathbb{P}(E)$, where E is locally free of rank n.

LEMMA 4.3. <u>Put</u> $\xi = c_1(\mathcal{O}_{\mathbb{P}(E)}(1)) \in A^{\cdot}\mathbb{P}(E)$, <u>and let</u>

$$\kappa : (A.Y)^n \to A.\mathbb{P}(E)$$

<u>be given by</u>

$$\kappa(y_0,\ldots,y_{n-1}) = \sum_{i=0}^{n-1} \xi^i \wedge \pi^* y_i$$

<u>where</u> $\pi : \mathbb{P}(E) \to Y$ <u>is the canonical morphism and</u> $\pi^* : A.Y \to A.\mathbb{P}(E)$ <u>the Gysin map</u>. <u>Then</u> κ <u>is an isomorphism of abelian groups</u>.

<u>Proof</u>: The usual proof from the non-singular case applies, in fact all one needs is the exact sequence

$$A.(X-U) \to A.X \to A.U \to 0$$

[F], 1.9, the property of "rational homotopy" [F], 4.3 as well as the observation

(4.3.1) $$\pi_*\Big(\sum_{i=0}^{n-1} \xi^i \wedge \pi^* y_i\Big) = y_{n-1}.$$

Indeed, one has

(4.3.2) $$\pi_*(\xi^i \wedge \pi^* y) = \begin{cases} 0 & \text{for} \quad i < n-1 \\ y & \text{for} \quad i = n-1 \end{cases}$$

Indeed, we may assume that $y = [V]$, V being a closed subvariety of Y. Restricting to V one may then assume that Y is reduced and irreducible and that $y = [Y]$. A dimension argument now yields the first line, while restriction to an open dense subset over which E is free yields the last line of (4.3.2).

We apply the lemma to $p : \mathbb{P}^N_Y \to Y$. Put $\dim(X/Y) = \dim X - \dim Y$ Then

(4.4)

$$i_*s_j(X/Y) = \sum \xi^{j+N-\dim(X/Y)-q} \wedge p^*\alpha_{j,q}$$

$$i_*\sigma_j(X/Y) = \sum \xi^{j+N-\dim(X/Y)-q} \wedge p^*\beta_{j,q}$$

$\alpha_{j,q}$ and $\beta_{j,q}$ are of codimension q, and the range of summation, where the α's and β's are defined, is

$$\dim Y \geq q \geq \text{Max}\ \{0, j - \dim(X/Y)\}$$

Remark. In particular, if $\dim Y = 0$ then there is only one ter in the sums, for $q = 0$. If $Y = \text{Spec}(k)$ then $\alpha_{j,0} = p_j$, $\beta_{j,0} = q_j$

Put $\eta = c_1(\mathcal{O}_{\underline{T}_Y}(1))$, where $\underline{T}_Y = P(\mathbb{P}^N_Y/Y)$, as before. The invariants $\alpha_{j,q}$ and $\beta_{j,q}$ can be described as follows:

PROPOSITION 4.5. *Put* $r(d,j) = d + \dim(X/Y) - j$ *and let* $P : \underline{T}_Y \to \mathbb{P}^N_Y$ *be the canonical morphism. Then*

(4.5.1) $$\alpha_{j,d} = P_*(\eta^{j+\dim(X/Y)-1} p^*\xi^{r(d,j)} \wedge [T(X/Y)])$$

(4.5.2) $$\beta_{j,d} = P_*(\eta^{j+\dim P(X/Y)-\dim X} p^*\xi^{r(d,j)} \wedge [P(X/Y)])$$

Proof: The two formulas being analogous, it suffices to show (4.5.1). First,

(4.5.3) $$i_*s_j(X/Y) = P_*(\eta^{j+\dim(X/Y)-1} \wedge [T(X/Y)])$$

which follows from the Projection Formula using the commutative diagram

$$\begin{array}{ccc} T(X/Y) & \xhookrightarrow{\ i'\ } & \underline{T}_Y \\ \downarrow{\scriptstyle \tau} & & \downarrow{\scriptstyle P} \\ X & \xhookrightarrow{\ i\ } & \mathbb{P}^N_Y \end{array}$$

together with $i'^*\eta = c_1(\mathcal{O}_{T(X/Y)}(1))$. Thus it suffices to show

$$(4.5.4) \qquad p_*(\xi^{r(d,j)} \wedge i_*s_j(X/Y)) = \alpha_{j,d}$$

By (4.4) we have

$$\xi^{r(d,j)} \wedge i_*s_j(X/Y) = \sum \xi^{N+d-q} \wedge p^*\alpha_{j,q}$$

and (4.5.4) follows by (4.3.1).

We next need the following lemma, which is an elementary special case of [F], 2.2. (4). Let E be a locally free sheaf on Y and $\phi : Y' \to Y$ a morphism. With $E' = \phi^*E$, one obtains the diagram

$$\begin{array}{ccc} \mathbb{P}(E') & \xrightarrow{\phi'} & \mathbb{P}(E) \\ \downarrow{\scriptstyle \pi'} & & \downarrow{\scriptstyle \pi} \\ Y' & \xrightarrow{\phi} & Y \end{array}$$

LEMMA 4.6. <u>If either Y is non-singular, or ϕ is flat, the two Gysin maps ϕ^* and ϕ'^* exist, and</u> $\pi'_*\phi'^* = \phi^*\pi_*$.

<u>Proof</u>: Let $x \in A.\mathbb{P}(E)$. Then $x = \sum_{i=0}^{n-1} \xi^i \wedge \pi^*y_i$, and thus

$$\phi'^*x = \sum_{i=0}^{n-1} \phi'^*(\xi^i \wedge \pi^*y_i) = \sum_{i=0}^{n-1} \phi'^*(\xi^i) \wedge \phi'^*\pi^*y_i$$

since the Gysin-map ϕ'^* is compatible with the contravariant structure of A. Now $\phi'^*(\xi) = \xi' = c_1(\mathcal{O}_{\mathbb{P}(E)}(1))$, and as the diagram of Gysin-maps,

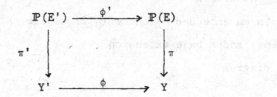

$$\begin{array}{ccc} A.\mathbb{P}(E') & \xleftarrow{\phi'^*} & A.(E) \\ {\scriptstyle \pi'^*}\uparrow & & \uparrow{\scriptstyle \pi^*} \\ A.Y' & \xleftarrow{\phi^*} & A.Y \end{array}$$

commutes, one obtains

$$(4.6.1) \qquad \phi'^*x = \sum_{i=0}^{n-1} \xi'^i \wedge \pi'^*\phi^*y_i \ .$$

By (4.3.1), $\pi'_*\phi'^*x = \phi^*y_i = \phi^*\pi_*x$, and the claim is proven.

We may now describe the behaviour of the invariants α and β under base change, at least in two important special cases. Fix the following notations:

Given an embedded Y-scheme (4.2), where X and Y have pure dimensions, and a base extension $Y' \to Y$, Y' of pure dimension. We get the diagram

$$\begin{array}{ccc} X_{Y'} = X' & \hookrightarrow & \mathbb{P}^N_{Y'} \\ & {\scriptstyle f'}\searrow \quad \swarrow{\scriptstyle p'} & \\ & Y' & \end{array}$$

Moreover, if F is a coherent sheaf on X whose support is of pure codimension q, define an element of $A.X'$ by

$$\mathrm{Tor}_+^{O_Y}(F, O_{Y'}) = \sum_{i=1}^{\infty} (-1)^{i-1} Z_d(\mathrm{Tor}_i^{O_Y}(F, O_{Y'}))$$

where $d = \dim X' - q$ (assuming that the sum is finite). Finally composing two diagrams like that of Lemma 4.6, we have

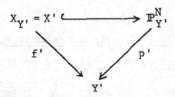

$$\begin{array}{ccc} \underline{T}_{Y'} & \xrightarrow{\phi'} & \underline{T}_{Y} \\ {\scriptstyle \pi'}\downarrow & & \downarrow{\scriptstyle \pi} \\ Y' & \xrightarrow{\phi} & Y \end{array}$$

For simplicity, let $e(j) = j + \dim(X/Y) - 1$.

THEOREM 4.7. *Assume either that* Y *is nonsingular or that* ϕ *is flat. Then*

$$\alpha'_{j,q} - \phi^*\alpha_{j,q} = \pi'_*(\phi'^*(\eta^{e(j)}p^*\xi^{r(q,j)}) \wedge \mathrm{Tor}_+^{O_Y}(O_{T(X/Y)}, O_{Y'}))$$

$$\beta'_{j,q} - \phi^*\beta_{j,q} = \pi'_*(\phi'^*(\eta^{e(j)}p^*\xi^{r(q,j)}) \wedge \mathrm{Tor}_+^{O_Y}(O_{P(X/Y)}, O_{Y'}))$$

where $\alpha_{j,q} = \alpha_{j,q}(X/Y,i)$, $\alpha'_{j,q} = \alpha_{j,q}(X'/Y',i')$, *etc.*

Moreover, if ϕ *is flat or if* f *is smooth, then the two expressions are zero.*

Proof: The last statement being obvious, and the formulas analogous, it suffices to prove the first one.

Put $T = T(X/Y)$, $T' = T(X'/Y')$, $\mathrm{Tor}_+ = \mathrm{Tor}_+^{O_Y}(O_T, O_{Y'})$. We have $T' = \phi'^{-1}(T)$, and

$$d = \dim T' = \dim T + \dim Y' - \dim Y = \dim \underline{T}_{Y'} - \mathrm{codim}\, T$$

hence

$$\phi'^*[T] = \sum_{i=0}^{\infty} (-1)^i Z_d(\mathrm{Tor}_i^{O_Y}(O_T, O_Y)),$$

see [S] and [F], 1.6 and 2.1. Now

$$[T'] = [O_T \otimes_{O_Y} O_{Y'}] = Z_d(O_T \otimes_{O_Y} O_{Y'}),$$

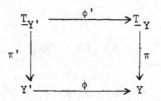

and hence

$$[T'] - \phi'^*[T] = \mathrm{Tor}_+$$

Let ξ', η' denote the elements in $A^{\cdot}\mathbb{P}^N_{Y'}$ and $A^{\cdot}\underline{T}_{Y'}$ which correspond to ξ, η. One finds

$$\eta'^{e(j)}p'^*\xi'^{r(q,j)} \wedge \mathrm{Tor}_+ = \eta'^{e(j)}p'^*\xi'^{r(q,j)} \wedge ([T'] - \phi'^*[T])$$

$$= \eta'^{e(j)}p'^*\xi'^{r(q,j)} \wedge [T'] - \phi'^*(\eta^{e(j)}p^*\xi^{r(q,j)} \wedge [T])$$

and thus

$$\pi'_*(\eta'^{e(j)}p'^*\xi'^{r(q,j)} \wedge \mathrm{Tor}_+) = \pi'_*(\eta'^{e(j)}p'^*\xi'^{r(q,j)} \wedge [T'])$$

$$- \quad \pi'_*(\phi'^*(\eta^{e(j)}p^*\xi^{r(q,j)} \wedge [T])) = \alpha'_{j,q} - \phi^*\alpha_{j,q}$$

by Proposition 4.5 and Lemma 4.6.

COROLLARY 4.7.1. (Joel Roberts) Let $f : X \to Y$ be a smooth, projective morphism with Y smooth and connected, X of pure dimension. Let $i : X \hookrightarrow \mathbb{P}^N_Y$ be a Y-embedding. Then the numbers $\mathrm{mult}_\tau(m)$ and $\mathrm{pinch}_\tau(m)$ are constant as τ runs through the geometr points of Y.

Proof: Since X_τ is smooth and of constant dimension, EGA IV, 17.5.2 and 6.1.2, one only needs to note that $\deg(s_i(X_\tau))$ is constan as τ varies.

Since for all j $\alpha_{j,0}(X/Y)$ is of codimension 0, $\alpha_{j,0}(X/Y) = m_{j,0}[Y]$. Now $\tau^*\alpha_{j,0}(X/Y) = m_{j,0}\tau^*[Y] = m_{j,0}[\mathrm{Spec}(K)]$, K being the field which corresponds to τ. By the theorem one now has $m_{j,0} = \deg(s_j(X_\tau))$, and the claim is immediate.

Remark: Roberts theorem is more general, in that he does not assume Y to be non-singular. On the other hand, our approach also applies to the case when f is not smooth.

References

[EGA] Grothendieck, A., (with the collaboration of Dieudonné, J.) Elements de géométrie algébrique. Chapters I - IV. Institut des Hautes Études Scientifiques, Publ. Math. 4, 8, 11, 17, 20, 24, 28, 32. Paris 1960-1967.

[F] Fulton, W., Rational equivalence on singular varities. Institut des Hautes Études Scientifiques, Publ. Math., 45 (1975), 147-167.

[F-M] Fulton, W., and MacPherson, R., Intersecting cycles on an algebraic variety. Preprint Series 1976/77, No. 14, Department of Mathematics, Aarhus Univ.

[H1] Holme, A., The notion of secant scheme for quasi-projective morphisms. Seminar reports, Matematisk Seminar, Univ. of Oslo, (1969).

[H2] ________, Formal embedding and projection theorems. Amer. Journ. of Math., 93 (1971), 527-571.

[H3] ________, A general embedding theorem in formal geometry. Compositio Math., 26 (1973), 41-68.

[H4] ________, Projections of non-singular projective varieties. J. Math. Kyoto Univ. 13 (1973), 301-322.

[H5] ________, An embedding-obstruction for algebraic varieties. Bull. Amer. Math Soc. 80 (1974), 932-934.

[H6] ________, Embedding obstruction for smooth, projective varieties I. To appear in Advances in Mathematics (1977), Preprint Series, Univ. of Bergen (1974).

[H7] ________, Embedding-obstruction for singular algebraic varieties in $\mathbb{P}^N$. Acta Math. 135 (1975), 155-185.

[H-R] ________, and Roberts, J., Pinch-points and multiple locus of generic projections of singular varieties. Preprint Series, Univ. of Bergen (1976).

[J] Johnson, K.W., Immersion and embedding of projective varieties. Thesis, Brown University, 1976.

[K] ________, The enumerative theory of singularities. Summer School on singularities, Oslo 1976. To appear on Wolters-Noordhoff Publishing, Groningen.

[Lak 1] Laksov, D., Some enumerative properties of secants to non-singular schemes. Math. Scand., 39 (1976), 171-190.

[Lak 2] ________, Secant bundles and Todd's formula for the double points of maps into $\mathbb{P}^n$. Preprint.

[Lak 3] ________, Residual intersections and Todd's formula for the double locus of a morphism. Preprint.

[Lau] Laudal, O.A., Sections of functors and the problem of lifting (deforming) algebraic structures I, II, III. Preprint series, Institute of Mathematics, Univ. of Oslo (1975).

[Lℓ 1] Lluis, E., Sur l'immersion des variétés algébriques, Ann. of Math., 62 (1955), 120-127.

[Lℓ 2] ________, De las singularidades que aparecen al proyectar variedades algebraicas. Bol. Soc. Mat. Mexicana, 1 (1956), 1-9.

[Lℓ 3] ________, Variedades algebraicas con ciertas condiciones en sus tangentes. Bol. Soc. Mat. Mexicana, (1962), 47-56.

[M] Manin, Yu.I., Lectures on the K-functor in algebraic geometry. Russian Mathematical Surveys, Vol. 24 (1969), 1-89.

[P.-S.] Peters, C.A.M., and Simonis, J., A secant formula. Quart. J. Math. Oxford, 27 (1976), 181-189.

[R1] Roberts, J., The variation of singular cycles in an algebraic family of morphisms, Trans. Amer. Math. Soc. 168 (1972), 153-164.

[R2] ___________, Singularity subschemes and generic projections. Trans. Amer. Math. Soc., 212 (1975), 229-268.

[R3] ___________, A stratification of the dual variety (Summary of results with indications of proof), preprint (1976).

[R4] ___________, Hypersurfaces with nonsingular normalization. Preprint (1977).

[S] Serre, J.-P., Algèbre Locale. Multiplicités. Springer Lecture Notes, Vol. 11 (1965).

DEPTH INEQUALITIES FOR COMPLEXES

Birger Iversen

Throughout this paper we consider bounded complexes $X^{\cdot}$ of finitdy generated modules over a local noetherian ring A. For an ideal I in we introduce the notion

$$\operatorname{depth}_I X^{\cdot} = \inf\{i \mid R^i\Gamma_I(X^{\cdot}) \neq 0\} - \sup\{i \mid H^i(X^{\cdot}) \neq 0\}$$

In case A is a regular ring we prove (amp $X^{\cdot} = \sup\{i \mid H^i(X^{\cdot}) \neq 0\}$ $\inf\{i \mid H^i(X^{\cdot}) \neq 0\}$)

$$\operatorname{depth}_I X^{\cdot} + \operatorname{amp} X^{\cdot} \leq \operatorname{depth}_I A$$

This is done by proving for complexes $X^{\cdot}$ and $Y^{\cdot}$ of free modules

$$\operatorname{amp} X^{\cdot} \otimes Y^{\cdot} \geq \operatorname{amp} X^{\cdot} + \operatorname{amp} Y^{\cdot}$$

The depth inequality above is worth conjecturing in the more general case where $V(I)$ is the support of a module of finite projective dimension. As is demonstrated it has as consequence the codimension conjecture of Auslander and the strong intersection conjecture of Peskine and Szpiro.

To support the conjecture and in order to be able to give examples the following lifting problem is considered: "Find a regular local ring R, a surjective morphisme $R \to A$ and a finitely generated R-module P such that $\operatorname{Supp} P \otimes_R A = V(I)$

and $Tor_i^R(P,A) = 0$ for $i > 0$." - We show that the depth inequality is valid whenever the lifting problem can be solved for the pair (A,I).

1° *I*-depth of a complex

Throughout this paper A denotes a noetherian local ring with maximal ideal m and residue field k. By a module is inderstood an A-module and by a complex is understood a complex of (A-) modules.

Lemma 1.1. Let $L^\cdot$ be a bounded above complex with finitely generated cohomology modules and $X^\cdot$ a bounded below complex of injective modules with finitly generated cohomology modules. If $H^\cdot(L^\cdot) \neq 0$ and $H^\cdot(X^\cdot) \neq 0$, then

$$H^\cdot(\mathrm{Hom}^\cdot(L^\cdot,X^\cdot)) \neq 0$$

Proof. Suppose $H^i(X^\cdot) \neq 0$, and choose $p \in \mathrm{Ass}\, H^i(X^\cdot)$. Put $d = \dim A/p$. It follows from [7], 2.6 that $\mathrm{Ext}^{i+d}(k,X^\cdot)$. In particular, with the notation of [7] $\mu(X^\cdot,t) \neq 0$. On the other hand we have [7], 1.2.

$$\mu(\mathrm{Hom}^\cdot(L^\cdot,X^\cdot),t) = \beta(L^\cdot,t)\mu(X^\cdot,t)$$

Q.E.D.

In the rest of this section I denotes an ideal in A contained in m.

Lemma 1.2. Let $X^{\cdot}$ be a bounded below complex with finitely generated cohomology modules. If $H^{\cdot}(X^{\cdot}) \neq 0$, then

$$\inf\{i \mid R^{i}\Gamma_{I}(X^{\cdot}) \neq 0\} = \inf\{i \mid \mathrm{Ext}^{i}(A/I, X^{\cdot}) \neq 0\}$$

For a finitely generated module P with $\mathrm{Supp}\, P = V(I)$, we have

$$\inf\{i \mid \mathrm{Ext}^{i}(P, X^{\cdot}) \neq 0\} = \inf\{i \mid \mathrm{Ext}^{i}(A/I, X^{\cdot}) \neq 0\}$$

Proof. Let $E^{\cdot}$ be a minimal injective resolution of $X^{\cdot}$, compare [7], the proof of 2.1. Note, that $\mathrm{Hom}^{\cdot}(A/I, E^{\cdot})$ is a minimal complex. Put

$$n = \inf\{i \mid \mathrm{Hom}(A/I, E^{i}) \neq 0\}$$

Note that

$$\mathrm{Hom}(A/I, E^{n}) \to \mathrm{Hom}(A/I, E^{n+1})$$

has a non trivial kernel since $\mathrm{Hom}(A/I, E^{\cdot})$ is a minimal complex. Thus

$$n = \inf\{i \mid \mathrm{Ext}^{i}(A/I, E^{\cdot}) \neq 0\}$$

It follows easily that $\Gamma_I(E^i) = 0$ for $i < n$ and $\mathrm{Ext}^n(A/I, E^\cdot)$ is a submodule of $R^n\Gamma_I(E^\cdot)$. – To prove the second part, note that

$$\mathrm{Hom}_A(P, E^\cdot) = \mathrm{Hom}_{A/I}(P, \mathrm{Hom}^\cdot(A/I, E^\cdot))$$

and consequently

$$\mathrm{Ext}^i_A(P, X^\cdot) = \mathrm{Ext}^i_{A/I}(P, \mathrm{Hom}^\cdot(A/I, E^\cdot))$$

In particular $\mathrm{Ext}^i(P, X^\cdot) = 0$ for $i < n$ and

$$\mathrm{Ext}^n(P, X^\cdot) = \mathrm{Hom}_{A/I}(P, \mathrm{Ext}^n(A/I, X^\cdot))$$

which is different from zero.

Q.E.D.

Definition 1.3. Let $X^\cdot$ be a bounded complex with finitely generated cohomology modules. If $H^\cdot(X^\cdot) \neq 0$, put

$$\mathrm{depth}_I X^\cdot = \inf\{i \mid R^i\Gamma_I(X^\cdot) \neq 0\} - \sup\{i \mid H^i(X^\cdot) \neq 0\}$$

Example 1.4. Let $X^\cdot$ be as above. If $V(I) \supseteq \mathrm{Supp}\, X^\cdot$, then

$$\mathrm{depth}_I X^\cdot = -\mathrm{amp}\, X^\cdot$$

Proposition 1.5. Let $X^\cdot$ be a bounded complex with finitely generated cohomology modules and $L^\cdot$ a bounded complex of finitely generated free modules. If $H^\cdot(X^\cdot) \neq 0$ and $H^\cdot(L^\cdot) \neq 0$, we have

$$\text{depth}_I X^\cdot \otimes L^\cdot \geq \text{depth}_I X^\cdot - \text{proj.amp}\, L^\cdot$$

Proof. Put $L^\cdot = \text{Hom}^\cdot(L^\cdot, A)$, and assume $X^\cdot$ is a bounded below complex og injective modules. We have isomorphisms of complexes

$$L^\cdot \otimes X^\cdot \xrightarrow{\sim} \text{Hom}^\cdot(L^{\cdot\vee}, X^\cdot), \quad \text{and}$$

$$\text{Hom}^\cdot(A/I, \text{Hom}^\cdot(L^{\cdot\vee}, X^\cdot)) \xrightarrow{\sim}$$

$$\text{Hom}^\cdot(A/I \otimes L^{\cdot\vee}, X^\cdot) \xrightarrow{\sim}$$

$$\text{Hom}^\cdot(A/I \otimes L^{\cdot\vee}, \text{Hom}^\cdot(A/I, X^\cdot))$$

Note that $\text{Hom}^\cdot(A/I, X^\cdot)$ is complex of injective A/I-modules, to conclude

$$\inf\{i \mid R^i\Gamma_I(\text{Hom}^\cdot(L^{\cdot\vee}, X^\cdot) \neq 0\} \geq$$

$$\inf\{i \mid R^i\Gamma_I(X^\cdot) \neq 0\} - \sup\{i \mid H^i(L^{\cdot\vee}) \neq 0\}$$

Combine this with

$$\sup\{i \mid H^i(X^\cdot \otimes L^\cdot) \neq 0\} =$$

$$\sup\{i \mid H^i(X^\cdot) \neq 0\} + \sup\{i \mid H^i(L^\cdot) \neq 0\}$$

to obtain

$$\text{depth}_I X^\cdot \otimes L^\cdot - \text{depth}_I X^\cdot \geq$$
$$- \sup\{i \mid H^i(L^\cdot) \neq 0\} - \sup\{i \mid H^i(L^{\cdot\vee}) \neq 0\}$$

Thus it will suffice to prove that

$$\text{proj.amp}\, L^{\cdot} = \sup\{i \mid H^{i}(L^{\cdot}) \neq 0\} + \sup\{j \mid H^{j}(L^{\vee}) \neq 0$$

which we leave to the reader.

Q.E.D.

Remark 1.6. In case $V(I) \supseteq \operatorname{Supp} L^{\cdot}$ we deduce from 1.5 and 1.4

$$\operatorname{depth}_I X^{\cdot} + \operatorname{amp} X^{\cdot} \otimes L^{\cdot} \leq \operatorname{proj\,amp} L^{\cdot}$$

which contains Hochster's "Tor inequality" [6], Theorem 1.

Corollary 1.7. Let $L^{\cdot}$ be a bounded complex of finitely generated free modules. If $H^{\cdot}(L^{\cdot}) \neq 0$. Then

$$\operatorname{depth}_I A \leq \text{proj.amp}\, L^{\cdot} + \operatorname{depth}_I L^{\cdot}$$

Example 1.8. Consider a complex

$$0 \to X_s \to X_{s-1} \to \ldots X_1 \to X_0 \to 0$$

such that for $i = 0,\ldots,s$ $(X_i \neq 0)$

1) $\operatorname{depth}_I X_i \geq i$

2) $H_j(X_{.}) = 0$ or $\operatorname{depth}_I H_j(X_{.}) = 0$.

From 1) we get from an easy spectral sequence argument, that depth $X. \geq 0$. From 2) we deduce that

$$\operatorname{depth}_I X. = -\operatorname{amp} X.$$

In conclusion $\operatorname{amp} X. = 0$. This is Peskine and Szpiro's celebrated acyclicity lemma [10], I.1.8.

Remark 1.9. Given a surjection $f: R \to A$ of local rings, an ideal J in R with $V(f(J)) = V(I)$. Then for a bounded complex $X^{\cdot}$ of R-modules with finitely cohomology modules and $H^{\cdot}(X^{\cdot}) \neq 0$ we have

$$\operatorname{depth}_J X^{\cdot}{}_{[f]} = \operatorname{depth}_I X^{\cdot}$$

as it follows from standard properties of local cohomology, [5].

2° The Buchsbaum-Eisenbud criterion, revised

In this section we shall give a very general example on the interrelationship between depth inequalities and amplityde inequalities. The example is based on the Buchsbaum-Eisenbud criterion in a form slightly different from the original [2]. The proof however is the same as that of the original.

Proposition 2.1. Consider a complex L. of finitely generated free modules

$$0 \to L_n \xrightarrow{d_n} L_{n-1} \xrightarrow{d_{n-1}} \dots L_1 \xrightarrow{d_1} L_0 \to 0$$

Let N be a finitely generated module such that

i) $d_p \otimes 1 : L_p \otimes N \to L_{p-1} \otimes N$
is different from zero for $p = 1, \dots, n$

ii) $H_p(L_{\bullet} \otimes N) = 0$ for $p = 1, \dots, n$

Then for $p = 1, \dots, n$

$$r_p = \sum_{i \geq 0} (-1)^i \operatorname{rk} L_{p+i} > 0$$

Proposition 2.2. Consider a complex of finitely generated free modules

$$0 \to L_n \xrightarrow{d_n} L_{n-1} \xrightarrow{d_{n-1}} \dots L_1 \xrightarrow{d_1} L_0 \to 0$$

such that for all $p = 1, \dots, n$

$$r_p = \sum_{i \geq 0} (-1)^i \operatorname{rk} L_{p+i} > 0$$

Let I_p denote the ideal generated by the r_p-minors of (a matrix for) d_p. Then for a finitely generated module $N \neq 0$, the following two conditions are equivalent

1) $H_p(L. \otimes N) = 0$ for $p \neq 0$.

2) For $p = 1, \dots, n$, $I_p = A$ or $\operatorname{depth}_{I_p} N \geq p$.

Corollary 2.3. With the notation and assumptions of 2.2 suppose in addition

$$V(I) = V(I_p), \quad p = 1, \dots, n$$

Then for any finitely generated module $N \neq 0$,

$$\operatorname{depth}_I N = \operatorname{proj.amp} L^\cdot - \operatorname{amp} L^\cdot \otimes N$$

(i.e. the inequality 1.6 is an equality).

Theorem 2.4. Let $P \neq 0$ be a finitely generated module. The following conditions are equivalent

i) For any bounded complex $L^\cdot$ of finitely generated free modules with $H^\cdot(L^\cdot) \neq 0$,

$$\operatorname{amp} P \otimes L^\cdot \geq \operatorname{amp} L^\cdot$$

ii) For any ideal $I \neq A$,

$$\operatorname{depth}_I P \leq \operatorname{depth}_I A$$

iii) Any P-regular sequence $\alpha_1, \dots, \alpha_s$ is an A-regular sequence.

Proof. That i) implies iii) follows by forming the Koszul complex on $\alpha_1, \dots, \alpha_s$. That iii) implies ii) is obvious. That ii) implies i) follows from Buchsbaum-Eisenbud criterion as follows: Let L. be represented

$$0 \to L_n \xrightarrow{d_n} L_{n-1} \xrightarrow{d_{n-1}} \dots L_1 \xrightarrow{d_1} L_0 \to 0$$

We may assume $\operatorname{amp} L. \otimes P = 0$, otherwise replace $L.$ by a truncation. Suppose first

$$d_i \otimes 1 : L_i \otimes P \to L_{i-1} \otimes P$$

is zero for some $i = 1, \ldots, n$. Then we conclude by Nakayamas lemma, that

$$0 \to L_n \to L_{n-1} \to \ldots L_i \to 0$$

is homotopic to zero. Thus by 2.1 we may assume

$$r_p = \sum_{i \geq 0} (-1)^i \operatorname{rk} L_{p+i} > 0$$

for $p = 1, \ldots, n$. We can now conclude by 2.2.

Q.E.D.

<u>Remark 2.5</u>. In case A is a regular local ring 2.3. iii follows easily from Serre's intersection inequality, compare [10], II.0.3. An alternative proof of 2.3, in this case has been given by Auslander [1] using "Tor-rigidity" [9]. A third proof is given in §4. - It is generally believed that the equivalent conditions of 2.3 hold in case P has a finite free resolution, Auslander's "zero-divisor conjecture". This conjecture has been proved in case A is equicharacteristic [10].

3° A strong intersection property

Throughout $P \neq 0$ denotes a module which admits a finite resolution $L^{\cdot}$ by finitely generated free modules. We let I be an ideal such that

$$V(I) = \operatorname{Supp} P$$

Theorem 3.1. Suppose A has a dualizing complex. Then, the following three inequalities holds simultaneously for all bounded complexes $X^{\cdot}$ with finitely generated cohomology modules and $H^{\cdot}(X^{\cdot}) \neq 0$ $(L^{\cdot\vee} = \operatorname{Hom}^{\cdot}(L^{\cdot}, A))$

i) $$\operatorname{depth}_I X^{\cdot} + \operatorname{amp} X^{\cdot} \leq \operatorname{depth}_I A$$

ii) $$\dim X^{\cdot} \otimes L^{\cdot} \geq \dim X^{\cdot} - \operatorname{depth}_I A$$

iii) $$\operatorname{amp} X^{\cdot} \otimes L^{\cdot\vee} \geq \operatorname{amp} X^{\cdot} + \operatorname{amp} L^{\cdot\vee}$$

Proof. Let us first prove the equivalence of i) and iii). To this we shall simply prove that

$$\operatorname{depth}_I A - \operatorname{depth}_I X^{\cdot} = \operatorname{amp} L^{\cdot\vee} \otimes X^{\cdot} - \operatorname{amp} L^{\cdot\vee}$$

We have (using 1.2)

$$\begin{aligned}\operatorname{amp} L^{\cdot\vee} \otimes X^{\cdot} &= \sup\{i \mid H^i(L^{\cdot\vee}) \neq 0\} + \sup\{i \mid H^i(X^{\cdot}) \neq 0\} \\ &\quad - \inf\{i \mid \operatorname{Ext}^i(P, X^{\cdot})\} \\ &= \sup\{i \mid H^i(L^{\cdot\vee}) \neq 0\} - \operatorname{depth} X^{\cdot} \\ &= \operatorname{amp} L^{\cdot\vee} + \inf\{i \mid \operatorname{Ext}^i(P, A) \neq 0\} - \operatorname{depth}_I X^{\cdot} \\ &= \operatorname{amp} L^{\cdot\vee} + \operatorname{depth}_I A - \operatorname{depth}_I X^{\cdot}.\end{aligned}$$

To prove the equivalence between ii) and iii) we shall prove the following formula ($D^\cdot$, dualizing complex for $X^\cdot$, $X^{\cdot D}$ the dual of $X^\cdot$ with respect to $D^\cdot$)

$$\dim X^\cdot \otimes L^\cdot - \dim X^\cdot =$$

$$\operatorname{amp} X^{\cdot D} \otimes L^{\cdot \vee} - \operatorname{amp} X^{\cdot D} - \operatorname{proj.amp} L^\cdot$$

To do so we shall freely use the results of [7], compare the proof of [7] 4.1.

$$\dim X^\cdot \otimes L^\cdot - \dim X^\cdot =$$

$$\operatorname{amp} L^{\vee\cdot} \otimes X^{\cdot D} - \operatorname{amp} X^{\cdot D} + \operatorname{depth} X^\cdot \otimes L^\cdot - \operatorname{depth} X^\cdot =$$

$$\operatorname{amp} L^{\cdot \vee} \otimes X^{\cdot D} - \operatorname{amp} X^{\cdot D} - \operatorname{proj.amp} L^\cdot$$

It will now suffice to prove that

$$\operatorname{amp} L^{\cdot \vee} = \operatorname{proj.amp} P - \operatorname{depth}_I A$$

which we leave to the reader to derive from 1.2.

Q.E.D.

Corollary 3.2. Suppose A has a dualizing complex. If P satisfies the equivalent conditions of 3.1, then

a) $\dim A/I + \operatorname{depth}_I A = \dim A$

b) For any finitely generated module $N \neq 0$ for which $N \otimes A/I$ has finite length, we have $\dim N \leq \operatorname{depth}_I A$

Proof. From 3.1.iii we get with $X^{\cdot} = A$, $\dim L^{\cdot} \geq \dim A - \operatorname{depth}_I A$, or $\dim A/I + \operatorname{depth}_I A \geq \dim A$.

The opposite inequality is wellknown to be generally valid, compare [10] I.4.8. - To prove b) note that the assumption can be interpreted $\dim \operatorname{Supp} L^{\cdot} \otimes N = 0$, in particular $\dim L^{\cdot} \otimes N = 0$. Thus we get

$$\dim N \leq \operatorname{depth}_I A$$

Q.E.D.

Remark 3.3. The formula in 3.2.a is known as Auslanders codimension conjecture [10], II.0.8, while 3.2.b is Peskine and Szpiros strong intersection conjecture [10], II.0.7.

Remark 3.4. In case $\operatorname{depth}_I A = \operatorname{proj.dim} P$ we have $\operatorname{amp} L^{\cdot \vee} = 0$. If moreover A is equicharacteristic it follows from [7], 3.2 that the depth inequality in 3.1 is valid.

4° Regular local rings

The purpose of this section is to prove a general amplitude inequality for an arbitrary regular local ring. This inequality was proved in the equicharacteristic case in [7].

<u>Theorem 4.1</u>. Let A be a regular local ring, $X^\cdot$ and $Y^\cdot$ bounded complexes of finitely generated free modules. If $H^\cdot(X^\cdot) \neq 0$ and $H^\cdot(Y^\cdot) \neq 0$, then

$$\operatorname{amp} X^\cdot \otimes Y^\cdot \geq \operatorname{amp} X^\cdot + \operatorname{amp} Y^\cdot$$

<u>Proof</u>. Let $X^\cdot$ and $Y^\cdot$ be normalized such that

$$H^0(X^\cdot) \neq 0, \; H^i(X^\cdot) = 0 \quad \text{for} \quad i < 0$$

$$H^0(Y^\cdot) \neq 0, \; H^i(Y^\cdot) = 0 \quad \text{for} \quad i < 0$$

Note that we have

$$\sup\{i \mid H^i(X^\cdot \otimes Y^\cdot) \neq 0\} = \sup\{i \mid H^i(X^\cdot) \neq 0\} + \sup\{i \mid H^i(Y^\cdot) \neq 0\}$$

Thus the amplitude inequality is equivalent to $H^i(X^\cdot \otimes Y^\cdot) \neq 0$ for some $i \leq 0$.

We shall now procede by induction on $\dim A$. The case $\dim A = 0$ follows from the Künneth formula.

Case 1, $\operatorname{Supp} H^0(X^\cdot) \cap \operatorname{Supp} H^0(Y^\cdot) \neq \{m\}$. Choose $p \in \operatorname{Supp} H^0(X^\cdot) \cap \operatorname{Supp} H^0(Y^\cdot)$ with $p \neq m$. Localize at p and notice that the normalization of $X^\cdot$ and $Y^\cdot$ has not been destroyed. By the induction hypotesis

$$H^i(X^\cdot_p \otimes Y^\cdot_p) \neq 0 \quad \text{for some} \quad i \leq 0.$$

Case 2, $\operatorname{Supp} H^0(X^\cdot) \cap \operatorname{Supp} H^0(Y^\cdot) = \{m\}$. By Serre's intersection theorem,

$$\dim H^0(X^\cdot) + \dim H^0(Y^\cdot) \leq \dim A$$

By [7] 2.4 we have

$$\dim X^{\cdot\vee} \leq \dim H^0(X^\cdot) \text{ and } \dim Y^{\cdot\vee} \leq \dim H^0(Y^\cdot)$$

and consequently

$$\dim X^{\cdot\vee} + \dim Y^{\cdot\vee} \leq \dim A$$

Using the results of [7] we get

$$\operatorname{amp} X^\cdot \otimes Y^\cdot - \operatorname{amp} X^\cdot - \operatorname{amp} Y^\cdot =$$

$$\dim X^{\cdot\vee} \otimes Y^{\cdot\vee} - \dim X^{\cdot\vee} - \dim Y^{\cdot\vee}$$

$$- \operatorname{depth} X^\cdot \otimes Y^\cdot + \operatorname{depth} X^\cdot + \operatorname{depth} Y^\cdot =$$

$$\dim X^{\cdot\vee} \otimes Y^{\cdot\vee} - \dim X^{\cdot\vee} - \dim Y^{\cdot\vee} + \dim A \geq$$

$$\dim A - \dim X^{\cdot\vee} - \dim Y^{\cdot\vee} \geq 0$$

Q.E.D.

<u>Corollary 4.2</u>. With the notation above

$$\dim A + \dim X^\cdot \otimes Y^\cdot \geq \dim X^\cdot + \dim Y^\cdot$$

Proof. This is simply the dual form of 4.1. Compare [7], 5.2.

Corollar 4.3. Let $I \neq A$ be an ideal in the regular local ring A. For a bounded complex of finitely generated modules with $H^{\cdot}(X^{\cdot}) \neq 0$,

$$\operatorname{depth}_I X^{\cdot} + \operatorname{amp} X^{\cdot} \leq \operatorname{depth}_I A$$

Proof. Follows from 3.1 and 4.1.

Q.E.D.

5° A lifting problem

Throughout this section $I \neq A$ denotes an ideal for which there exists a finitely generated module Q of finite projective dimension with $\operatorname{Supp} Q = V(I)$. - Consider the following lifting problem

5.1 Find a regular local ring R, a surjective morphism $R \to A$ and a finitly generated R-module P, with

i) $\operatorname{Tor}_i^R(P, A) = 0, \quad i > 0$

ii) $\operatorname{Supp} P \otimes_R A = V(I)$

Theorem 5.2. If the lifting problem 5.1 can be solved for (A,I), then for any bounded complex $X^{\cdot}$ of finitly generated modules and $H^{\cdot}(X^{\cdot}) \neq 0$, we have

$$\operatorname{depth}_I X^{\cdot} + \operatorname{amp} X^{\cdot} \leq \operatorname{depth}_I A$$

Proof. Let $R, R \to A, P$ solve the lifting problem and put $J = \operatorname{Ann} P$. We are first going to prove that

$$\operatorname{depth}_J R = \operatorname{depth}_I A$$

From 2.4 (or 4.3) we get

$$\operatorname{depth}_J R \geq \operatorname{depth}_J A \quad (= \operatorname{depth}_I A)$$

Consider a resolution $L.$ of P by finitely generated free modules

$$0 \to L_n \xrightarrow{d_n} L_{n-1} \cdots \to L_2 \xrightarrow{d_2} L_1 \xrightarrow{d_1} L_0 \to 0$$

As is well known $\operatorname{depth}_J A = 0$ if and only if $\chi(P) = \Sigma(-1)^i \operatorname{rk} L_i \neq 0$. Thus $\operatorname{depth}_J R = 0$ and $\operatorname{depth}_I A = 0$ simultaneously. Thus we may assume $\chi(P) = 0$. Put $r_p = \Sigma_{i \geq 0}(-1)^i \operatorname{rk} L_{p+i}$ and let I_p denote the ideal generated by the r_p-minors of d_p. Recall [3], Thm. 3.1, there is a nonzero divisor $\alpha \in A$ such that $(V(I_1)=V(J))$

$$I_1 = (\alpha) I_2$$

(The ideal (α) is the "Mac Rae invariant of P" or the "determinant of $0 \to L.$" in the sense of Knudsen and Mumford [8]).

We have $\text{depth}_J R = 1$ if and only if $(\alpha) \neq R$. It is cleare that the image of (α) in A is the Mac Rae invariant of $P \otimes_R A$. Thus $\text{depth}_J R = 1$ implies $\text{depth}_I A = 1$. Suppose $\text{depth}_J R = k > 1$. Then by [3], Theorem 2.1. We have

$$V(I_p) = V(I) \quad \text{for} \quad p \leq k$$

On the other hand since $H^i(L^\cdot \otimes_R A) = 0$ for $i > 0$ we have by the Buchsbaum-Eisenbud criterion 2.2

$$\text{depth}_{I_p} A \geq p, \quad p = 1, \ldots, n$$

in particular $(V(I_k) = V(J))$

$$\text{depth}_J A \geq k = \text{depth}_J R$$

To conclude the proof consider a complex $X^\cdot$ of A-modules, as in the statement of the proposition. By 4.3

$$\text{depth}_J X^\cdot + \text{amp}\, X^\cdot \leq \text{depth}_J R$$

and whence by the previous result and 1.9

$$\text{depth}_I X^\cdot + \text{amp}\, X^\cdot \leq \text{depth}_I A$$

Q.E.D.

Remark 5.3. We refer the reader to [10] I.2 for a counter example to the general problem of lifting a module of finite projective dimension.

Example 5.4. There are a number of examples of canonical resolutions of minors of various types of matrices which immediately may be lifted by lifting the matrices themselves, see [3] and [4]. The examples (all?) leads to perfect ideals for which the issue already is settled in the equicharacteristic case, compare 3.4. - To get a non perfect example, suppose $\alpha, \beta, \gamma, \delta$ is a regular sequence in A, then the ideal $(\alpha\delta, \beta\gamma, \alpha\gamma-\beta\delta)$ has a length 3 resolution (Kaplansky), see [3] 11.2, which may be lifted anywhere.

REFERENCES

[1] Auslander, M., Modules over unramified regular local rings, Ill. J.Math. 5 (1961) p. 631-645,

[2] Buchsbaum, D.A. and Eisenbud, D., What makes a complex exact? J. Algebra 25 (1973) p. 49-58,

[3] Buchsbaum, D.A., Eisenbud, D., Some structure theorems for finite free resolutions. Advances in mathematics 12 (1974) p. 84-139,

[4] Buchsbaum, D.A., Eisenbud, D., Algebra structures for finite free resolutions, and some structure theorems for ideals of codimension 3. Amer. J. Math.99 (1977) p. 447-485,

[5] Grothendieck, A., (Notes by R. Hartshorne), Local Cohomology. Lecture Notes in Mathematics 41. Springer Verlag, Berlin 1967,

[6] Hochster, M., Grade sensitive modules and perfect modules. Proc. London Math. Soc. (3) 29 (1974) p.55-76,

[7] Iversen, B., Amplitude inequalities for complexes. to appear in Ann.scient. Éc. Norm. Sup. X (1977),

[8] Knudsen, F., and Mumford, D., The projectivity of the moduli space of stable courves I. Math. Scand. 39 (1976) p. 19-55.

[9] Lichtenbaum, S., On the vanishing of Tor in regular local rings. Ill. J. Math. 10 (1966) p. 220-226,

[10] Peskine, C., and Szpiro, L., Dimension projective finie et cohomologie locale. Publ. Math. I.H.E.S. no. 42 (1973) p. 323-395.

Aarhus Universitet

October 1977
/ET

A generalized trisecant lemma

by

O.A. Laudal

Introduction. Let k be a field and consider a closed subscheme X of $\mathbb{P}_k^N$.

One obvious way of studying X is to pick a family of simple subscheme $\{Y\}$ of $\mathbb{P}_k^N$ and study the corresponding family of intersections $\{X \cap Y\}$.

This is a method which has been used since the very beginning of algebraic geometry.

The first non-trivial case, when X is a non-singular curve of $\mathbb{P}_k^3$ and the family $\{Y\}$ is the family of lines, leads to the study of the set of lines L with $X \cap L$ of given type.

It has been known for a long time that in the "general case", i.e. when X is not "special", there is a 2 dimensional family of lines intersecting X in 2 points, and there is a 1 dimensional family of trisecants, i.e. lines intersecting X in 3 points.

The real problem is, of course, what should be ment by "general case" or "special".

The classical answer is given by the trisecant lemma stating that the family of trisecants is always of dimension 1 unless X is a plane curve of degree 3, in which case it is trivially seen to be of dimension 2.

The general situation is, of course, more complex.

We shall have to work quite a lot to see what kind of results one should expect in the "general case". And, at the moment, the results analogous to the trisecant lemma, are scarce.

We shall start by making the setting more precise:

Pick any pair of closed subschemes $\mathbb{h}, \mathbb{h}_o$ of $\mathrm{Hilb}_{\mathbb{P}^N}$ and consider the set

$$I = I(X; \mathbb{h}, \mathbb{h}_o) = \{Y \in \mathbb{h} \mid Z = X \cap Y \in \mathbb{h}_o\}$$

One easily proves that I has a natural scheme-structure. Moreover, assuming $\mathbb{h}$ is the irreducible component of a complete intersection, minimally contained in a linear subspace of dimension l of $\mathbb{P}^N_k$, we shall define a natural rational morphism

$$\mathfrak{g} : I \longrightarrow \mathrm{Grass}(l+1, N+1)$$

regular on the open subscheme I_o of I corresponding to the complete intersections. Using the deformation theory developed in [La 2] we shall study the formalization of $\mathfrak{g}$ at a closed point Y of I. In particular we shall compute the imbedding dimensions, and we shall give sufficient conditions for non-singularity of I at Y.

Finally we shall concentrate on the following reasonably well understood special cases.

§3. The theory of r-secants. This is the case corresponding to $\mathbb{h} = \mathrm{Grass}(2, N+1) \subseteq \mathrm{Hilb}_{\mathbb{P}^N}$ and $\mathbb{h}_o = \mathrm{Hilb}^r_{\mathbb{P}^N}$, the subscheme of $\mathrm{Hilb}_{\mathbb{P}^N}$ parametrizing finite closed subschemes of length r.

§4. The generalized r-secants of a curve in $\mathbb{P}^3$. This case corresponds to $N = 3$, $\dim X = 1$, $\mathbb{h}$ being the subscheme of $\mathrm{Hilb}_{\mathbb{P}^N}$ parametrizing the plane curves of degree n, and $\mathbb{h}_o = \mathrm{Hilb}^r_{\mathbb{P}^N}$.

§3 will include the classical results on the dimension of the family of r-secants, and in particular a new proof of the trisecant lemma.

The main result of §4 has the following Corollary:

Suppose the generic hyperplane section of a curve X in $\mathbb{P}^3_k$ contains a finite subscheme Z sitting on a (plane) curve of degree n but on no more than 3 curves of degree $(n+1)$, then X is contained in

a surface of degree n.
Since any 3 different colinear points in a plane are contained in exactly 3 conics, this result is an immediate generalization of the classical trisecant lemma.

The ideas, and some of the results of this paper have appeared in two preprints, No 24(1975) and No 4(1977) of the Preprint Series of the University of Oslo. Unfortunately the lemma (5.2.12) of the first one is erroneous, making the statements (5.2.13) and (5.2.14) not proved. (As we shall see they are in fact false.)

Notations.

A composition of morphisms $\cdot \xrightarrow{\varphi} \cdot \xrightarrow{\psi} \cdot$ in a category will be denoted by $\varphi\psi$ in accordence with the notations of [La 2].

$\underline{\text{sch}}/k$ is the category of k-schemes.

$k/\underline{\text{sch}}/k$ is the category of pointed k-schemes.

Functors and categories are denoted by underlined letters.

Given a scheme Y, the functor represented by Y is written: $\underline{Y} : \underline{\text{sch}}/k \to \underline{\text{sets}}$.

$\mathbb{P}^N_k$ is the projective N-space.

$\text{Hilb}_{\mathbb{P}^N}$ is the Hilbert scheme of $\mathbb{P}^N_k$ parametrizing the closed subschemes of $\mathbb{P}^N_k$.

Grass(m,n) is the Grassmanian parametrizing the m-dimensional linear subspaces of A^n, the affine n-space.

Commutative diagrams will be indicated by a small circle, cartesian diagrams with a little square.

§1. Construction of the scheme $I(X;h,h_o)$ and the morphism $\mathfrak{g}$.

Let X be a closed subscheme of $\mathbb{P}^N_k$. Pick two closed subschemes of the Hilbert scheme $\text{Hilb}_{\mathbb{P}^N}$, $\mathbb{h}$ and $\mathbb{h}_o$, and consider the functor

$$\underline{I} = \underline{I}(X;\, \mathbb{H},\, \mathbb{H}_o) : \underline{sch/k} \longrightarrow \underline{sets}$$

defined by:

$$\underline{I}(S) = \{Y_S \to \mathbb{P}^N\times S \mid (Y_S \to \mathbb{P}^N\times S) \in \underline{\mathbb{H}}(S),\ (Y_S \underset{\mathbb{P}^N\times S}{\times} (X\times S) \to \mathbb{P}^N\times S) \in \underline{\mathbb{H}}_o(S)\}$$

By definition $\underline{I}$ is a subfunctor of $\underline{\mathbb{H}}$ and we shall show that I is represented by a disjoint union of disjoint locally closed subschemes of $\mathbb{H}$. In fact let $\rho : \tilde{\mathbb{H}} \to \mathbb{H}$ and $\rho_o : \tilde{\mathbb{H}}_o \to \mathbb{H}_o$ be the universal families and consider the diagram

$$\begin{array}{ccc}
\mathbb{P}^N\times\mathbb{H} & \longleftarrow & X\times\mathbb{H} \\
\uparrow & & \uparrow \\
\tilde{\mathbb{H}} & \longleftarrow & \tilde{Z} = \tilde{\mathbb{H}} \underset{\mathbb{P}^N\times\mathbb{H}}{\times} (X\times\mathbb{H}) \\
& {\scriptstyle\rho}\searrow \quad \swarrow{\scriptstyle\eta} & \\
& \mathbb{H} &
\end{array}$$

Let $\{I'_\nu\}_\nu$ be a flattening stratification of $\mathbb{H}$ corresponding to $O_{\tilde{Z}}$ as an $O_{\mathbb{P}^N\times\mathbb{H}}$-Module. Let $\tilde{Z}'_\nu = \tilde{Z} \underset{\mathbb{H}}{\times} I'_\nu$, then $\tilde{Z}'_\nu \xrightarrow{\eta'_\nu} I'_\nu$ is flat and thus corresponds to a unique morphism $I'_\nu \to \mathrm{Hilb}_{\mathbb{P}^N}$. Put $I_\nu = I'_\nu \underset{\mathrm{Hilb}_{\mathbb{P}^N}}{\times} \mathbb{H}_o$. Then we claim that $I = \coprod_\nu I_\nu$ represents $\underline{I}$. By definition of $\underline{I}$ an element $(Y_S \to \mathbb{P}^N\times S)$ of $\underline{I}(S)$ corresponds to unique morphisms

$$\varphi : S \to \mathbb{H} \qquad\qquad \psi : S \to \mathbb{H}_o$$

such that the following diagrams exist

$$\begin{array}{ccc}
Y_S & \longrightarrow & \tilde{\mathbb{H}} \\
\downarrow & \square & \downarrow \\
S & \xrightarrow{\ \varphi\ } & \mathbb{H}
\end{array}
\qquad\qquad
\begin{array}{ccc}
Y_S \underset{\mathbb{P}^N\times S}{\times} (X\times S) & \longrightarrow & \tilde{\mathbb{H}}_o \\
\theta\downarrow & \square & \downarrow \\
S & \xrightarrow{\ \psi\ } & \mathbb{H}
\end{array}$$

The first one induces a cartesian diagram:

$$\begin{array}{ccc}
Y_S \underset{\mathbb{P}^N\times S}{\times} (X\times S) & \longrightarrow & \tilde{\mathbb{H}} \underset{\mathbb{P}^N\times\mathbb{H}}{\times} (X\times\mathbb{H}) = \tilde{Z} \\
\theta\downarrow & \square & \downarrow \\
S & \xrightarrow{\ \varphi\ } & \mathbb{H}
\end{array}$$

Since θ is flat φ factors through $\coprod_\nu I'_\nu$ giving birth to cartesian digrams

$$\begin{array}{ccccc} Y_S \underset{\mathbb{P}^N \times S}{\times} (X \times S) & \longrightarrow & \coprod_\nu \tilde{Z}'_\nu & \longrightarrow & \tilde{Z} \\ \theta \downarrow & \square & \downarrow & \square & \downarrow \\ S & \xrightarrow[\varphi']{} & \coprod_\nu I'_\nu & \xrightarrow[\varphi'']{} & \tilde{\mathbb{h}} \end{array}$$

with $\varphi'\varphi'' = \varphi$.

Moreover there is a unique cartesian diagram

$$\begin{array}{ccc} \coprod_\nu \tilde{Z}'_\nu & \longrightarrow & \widetilde{\mathrm{Hilb}}_{\mathbb{P}^N} \\ \downarrow & & \downarrow \\ \coprod_\nu I'_\nu & \xrightarrow[\psi'']{} & \mathrm{Hilb}_{\mathbb{P}^N} \end{array}$$

proving that $\psi = \varphi'\psi''$. It follows that φ' factors through $\coprod_\nu I_\nu = \psi''^{-1}(\mathbb{h}_o)$. We have therefore produced a unique morphism $\varphi_o : S \to \coprod_\nu I_\nu$. This defines a morphism of functors $\underline{I} \to \mathrm{Mor}(-, I)$ which is easily seen to be an isomorphism.

<u>Remark</u>. Given a Hilbert-polynomial P, suppose $\mathbb{h}_o \subseteq \mathrm{Hilb}^P_{\mathbb{P}^N}$, then $I = I_\nu$ for some ν, corresponding to the polynomial P. Thus I i a locally closed subscheme of $\mathbb{h}$.

Let $\pi : \tilde{I} \to I$ be the universal family and consider the diagrams

$$\begin{array}{ccc} \tilde{I} & \longrightarrow & \tilde{\mathbb{h}} \\ \pi\downarrow & & \downarrow \\ I & \longrightarrow & \mathbb{h} \end{array} \qquad \begin{array}{ccccc} \tilde{I} & \longrightarrow & \mathbb{P}^N_k \times I & \longrightarrow & \mathbb{P}^N_k \\ \pi\downarrow & & & & \\ I & & & & \end{array}$$

Let $\mathrm{Spec}(A) \subseteq I$ be an open affine subscheme, then the imbedding $i_A : \pi^{-1}(\mathrm{Spec}(A)) \to \mathbb{P}^N_k \otimes_k A$ is defined by a homogenous ideal C in th graded A-algebra $A[X_o, \ldots, X_N]$. By definition $\mathrm{Spec}(A[X_o, \ldots, X_N]/C)$ is the projecting cone of the subscheme $\pi^{-1}(\mathrm{Spec}(A))$ of $\mathbb{P}^N_k \otimes_k A$. Suppose the fibers of π at the points of $\mathrm{Spec}(A)$ corresponds to complete intersections of $\mathbb{P}^N_k$, then by ([E], 1.), the A-algebra $A[X_o, \ldots, X_N]/C$ is flat. Moreover $\mathrm{Spec}(A[X_o, \ldots, X_N]/C) \otimes_A \mathbb{k}_y$ is the projecting cone of $Y = \pi^{-1}(y)$ for all $y \in \mathrm{Spec}(A)$. Since the subse

I_o of I corresponding to complete intersections is open, we find by gluing the affine pieces that the projecting cone of $\tilde{I}_o = \pi^{-1}(I_o) \subseteq \mathbb{P}^N \times I_o$ is flat over I_o and specializes to the projecting cone of the fibers of π.

For every affine piece above, let $C(1)$ be the ideal of $A[X_o,\dots,X_N]$ generated by C_1, the elements of degree one of C. Since C_1 has locally constant rank on $\mathrm{Spec}(A)$, the graded A-algebra

$$A[X_o,\dots,X_N]/C(1)$$

is flat. Moreover $\mathrm{Proj}(A[X_o,\dots,X_N]/C(1)) \subseteq \mathbb{P}^N \times \mathrm{Spec}(A)$ is the linear span of $\pi^{-1}(\mathrm{Spec}(A))$. There correspond diagrams

$$\begin{array}{ccc} \pi^{-1}(\mathrm{Spec}(A)) = \mathrm{Proj}(^{A[X_o,\dots,X_N]}/C) & \subseteq \mathrm{Proj}(^{A[X_o,\dots,X_N]}/C(1)) & \subseteq \mathbb{P}^N \times \mathrm{Spec}(A) \\ \downarrow & \downarrow & \\ \mathrm{Spec}(A) & = \quad \mathrm{Spec}(A) & \end{array}$$

which pieces together to form a diagram

$$\begin{array}{ccccc} \tilde{I}_o & \hookrightarrow & \tilde{L} & \hookrightarrow & \mathbb{P}^N_k \times I_o \\ \pi_o\downarrow & & \downarrow\lambda & & \\ I_o & = & I_o & & \end{array}$$

in which π_o and λ are flat.

If $\mathbb{h}$ is the irreducible component (in $\mathrm{Hilb}_{\mathbb{P}^N}$) of a complete intersection Y, the diagram above defines a morphism

$$\mathfrak{g} : I_o \longrightarrow \mathrm{Grass}(l+1,N+1)$$

where l is the dimension of the linear span of Y in $\mathbb{P}^N_k$.

In fact, the condition implies that the dimension of the fibres of λ is constant, equal to l. Therefore mapping an element y of I_o onto the l-dimensional linear subspace $\lambda^{-1}(y) \subseteq \mathbb{P}^N_k$, defines a map $I_o \to \mathrm{Grass}(l+1,N+1)$ which, since λ is flat, is a morphism of k-schemes.

Pick a rational point $L \in \mathrm{Grass}(l+1,N+1)$, then the fibre of $\mathfrak{g}$ at

L, $g^{-1}(L)$ represents the functor

$$\underline{g^{-1}}(L) : \underline{sch}/k \longrightarrow \underline{sets}$$

defined by:

$$\underline{g^{-1}}(L)(S) = \{Y_S \hookrightarrow L\times S \subseteq \mathbb{P}^N\times S \mid (Y_S \hookrightarrow \mathbb{P}^N\times S)\in \underline{I}_o(S)\}$$

Given a rational point y of I_o corresponding to $Y = \pi^{-1}(y) \subseteq \mathbb{P}^N$ the fibre-functors of $\underline{g^{-1}}(g(y))$, $\underline{I}_o$ and $\underline{G} = \underline{\text{Grass}}(l+1,N+1)$ at the points y, y and $L = g(y)$ respectively, are defined on the category $k/\underline{sch}/k$ of pointed k-schemes.

They are, respectively

$$\text{Fib}_y\, \underline{g^{-1}}(L) : k/\underline{sch}/k \longrightarrow \underline{sets}$$

with

$$\text{Fib}_y(\underline{g^{-1}}(L))(*\to S) = \left\{ \begin{array}{ccc} Y_S & \to & L\times S \subseteq \mathbb{P}^N\times S \\ \uparrow & \square & \uparrow \\ Y & \to & L \subseteq \mathbb{P}^N \end{array} \middle| (Y_S \to \mathbb{P}^N\times S)\in \underline{g^{-1}}(L)(S) \right\},$$

$$\text{Fib}_y\, I_o : k/\underline{sch}/k \longrightarrow \underline{sets}$$

with

$$\text{Fib}_y\, \underline{I}_o(*\to S) = \left\{ \begin{array}{ccc} Y_S & \to & \mathbb{P}^N\times S \\ \uparrow & \square & \uparrow \\ Y & \to & \mathbb{P}^N \end{array} \middle| (Y_S \to \mathbb{P}^N\times S)\in \underline{I}_o(S) \right\}$$

and

$$\text{Fib}_L\, \underline{G} : k/\underline{sch}/k \longrightarrow \underline{sets}$$

given by

$$\text{Fib}_L\, \underline{G}(*\to S) = \left\{ \begin{array}{ccc} L_S & \to & \mathbb{P}^N\times S \\ \uparrow & \square & \uparrow \\ L & \to & \mathbb{P}^N \end{array} \middle| (L_S \to \mathbb{P}^N\times S)\in \underline{G}(S) \right\}$$

Let $\underline{l}$ be the full subcategory of $k/\underline{sch}/k$ given by the affine schemes $\text{Spec}(R)$ where R runs through the artinian local k-algebr[as] with residue field k. Then the restrictions of $\text{Fib}_y\, \underline{g^{-1}}(L)$, $\text{Fib}_y\, \underline{I}[_o]$ and $\text{Fib}_L\, \underline{G}$ to $\underline{l}$ are prorepresented by

$$\hat{O}_{g^{-1}(L),y}\ ,\ \hat{O}_{I_o,y} \text{ and } \hat{O}_{G,L} \quad \text{respectively.}$$

Moreover, the morphism g induces comorphisms

$$\hat{O}_{g^{-1}(L),y} \xleftarrow{j^*} \hat{O}_{I_o,y} \xleftarrow{g^*} \hat{O}_{G,L} .$$

The purpose of the next paragraph is to study these morphisms, and in particular, study the complete local k-algebra $\hat{O}_{I_o,y}$. This we shall do by studying the corresponding functors and the morphisms among those functors, using the machinery of [La 2].

§ 2. Deformation theory.

Consider a commutative diagram of closed imbeddings of k-schemes

$$(*)\qquad \begin{array}{ccccc} X & \xrightarrow{f} & \mathbb{P}^N & \xleftarrow{g} & Y \\ & \nwarrow^{\alpha} & \uparrow{\scriptstyle h} & \nearrow_{\beta} & \\ & & Z & & \end{array}$$

A deformation of this diagram, leaving f fixed, is a pointed k-scheme $* \to S$ and a commutative diagram of flat S-schemes

$$\begin{array}{ccccc} X\times S & \xrightarrow{f\times 1_S} & \mathbb{P}^N\times S & \xleftarrow{g'} & Y_S \\ & \nwarrow^{\alpha'} & \uparrow{\scriptstyle h'} & \nearrow_{\beta'} & \\ & & Z_S & & \end{array}$$

together with morphisms of k-schemes

$$Y \to Y_S \quad\text{and}\quad Z \to Z_S$$

making the following diagrams cartesian,

$$\begin{array}{ccc} \mathbb{P}^N\times S & \xleftarrow{g'} & Y_S \\ \uparrow & & \uparrow \\ \mathbb{P}^N & \xleftarrow[g]{} & Y \end{array} \qquad\qquad \begin{array}{ccc} \mathbb{P}^N\times S & \longleftarrow & \mathbb{P}^N \\ \uparrow{\scriptstyle h'} & & \uparrow{\scriptstyle h} \\ Z_S & \longleftarrow & Z \end{array}$$

For given $* \to S$, let $\mathrm{Def}_1(* \to S)$ be the set of all such deformations. This defines a functor

$$\mathrm{Def}_1 : k/\underline{sch}/k \to \underline{sets} .$$

If $Z = X \cap Y$ and if $\mathbb{h}$ and $\mathbb{h}_o$ contain the component of Y, respectively Z, in $\mathrm{Hilb}_{\mathbb{P}^N}$, the restrictions of $\mathrm{Fib}_Y \underline{I}$ and Def_1 to the subcategory $\underline{1}$ of $k/\underline{sch}/k$ coincide. To see this, observe that with the notations above there is a diagram

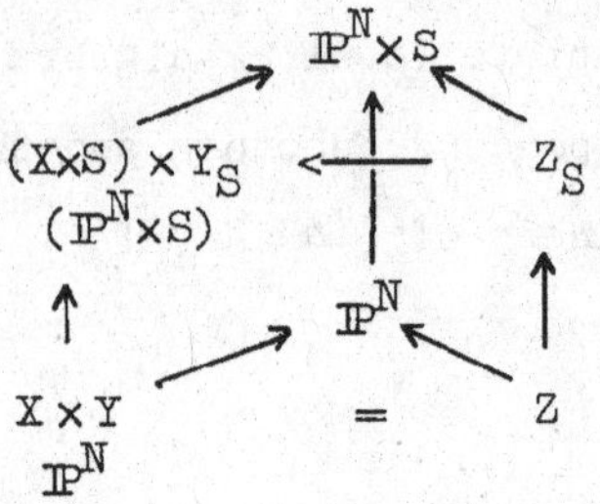

For $S = \mathrm{Spec}(R)$ an object of $\underline{1}$, the closed imbedding $(X \times S) \underset{\mathbb{P}^N \times S}{\times} Y_S \leftarrow Z_S$ must necessarily be an isomorphism, proving that

$$\mathrm{Def}_1 | \underline{1} = \mathrm{Fib}_Y \underline{I} | \underline{1} .$$

In the same way we may identify the restrictions to $\underline{1}$ of $\mathrm{Fib}_Y \underline{g}^{-1}(\underline{g}(Y$ and the deformation functor $\underline{\underline{\mathrm{Def}}}_o$ corresponding to the deformations o the diagram

(**)

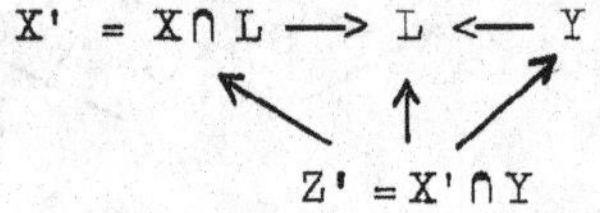

leaving X' fixed.

Finally, the fiberfunctor $\mathrm{Fib}_L \underline{G}$ and the deformation functor $\underline{\mathrm{Def}}_2$ corresponding to the diagram

(***) $$\mathbb{P}^N \longleftarrow L$$

coincide on $\underline{1}$.

Thus, studying $\mathbb{j}^*$ and $\mathbb{g}^*$ above, is equivalent to studying the restrictions to $\underline{1}$ of the morphisms $\mathrm{Def}_o \underset{\mathbb{j}}{\Rightarrow} \mathrm{Def}_1 \underset{\mathbb{g}}{\Rightarrow} \mathrm{Def}_2$.

This is where deformation theory enters. Using the general set-up of [La2] we shall be able to "compute" the hulls of the deformation functors Def_i, $i = 0,1,2$, i.e. the complete local rings

$$\hat{O}_{g^{-1}(L),Y}\ ,\quad \hat{O}_{I_o,Y} \quad \text{and} \quad \hat{O}_{G,L}$$

together with the morphisms j^* and g^*.
In fact, the main theorem of [La2] claims the following:
Given any small category $\underline{D}$ of morphisms of k-schemes, and any subcategory $\underline{D}_o$, then the corresponding deformation functor

$$\mathrm{Def}(\underline{D}/\underline{D}_o) : \underline{l} \longrightarrow \underline{\mathrm{sets}}$$

defined by (see [La2] 4.2):

$$\mathrm{Def}(\underline{D}/\underline{D}_o)(* \to S) = \{\text{deformations of } \underline{D} \text{ to } S \text{ , trivial on } \underline{D}_o\}/\sim$$

has a profinite hull H whenever the corresponding algebraic cohomology groups $A^i_{\underline{D}_o}(\underline{D},O_{\underline{D}})$ for $i = 1,2$ are of finite demension as k-vectorspaces. Moreover H is given in terms of the total obstruction morphism

$$o : \mathrm{Sym}_k(A^2_{\underline{D}_o}(\underline{D},O_{\underline{D}})^*)^\wedge \longrightarrow \mathrm{Sym}_k(A^1_{\underline{D}_o}(\underline{D},O_{\underline{D}})^*)^\wedge$$

by

$$H = \mathrm{Sym}_k(A^1_{\underline{D}_o}(\underline{D};O_{\underline{D}})^*) \underset{\mathrm{Sym}_k(A^2_{\underline{D}_o}(\underline{D},O_{\underline{D}})^*)^\wedge}{\hat{\otimes}} k$$

This needs some explanation. We shall concentrate on examples relevant to our special needs.
Let $\underline{D}$ be the category of k-morphisms corresponding to the diagram (*) above. The objects of $\underline{D}$ are f, g and h, the nontrivial morphisms being the diagrams

$$(\alpha,1) : h \to f : \quad \begin{array}{ccc} \mathbb{P}^N & \xrightarrow{1} & \mathbb{P}^N \\ h\uparrow & & \uparrow f \\ Z & \xrightarrow[\alpha]{} & X \end{array}$$

$$(\beta,1) : h \to f : \quad \begin{array}{ccc} \mathbb{P}^N & \xrightarrow{1} & \mathbb{P}^N \\ h\uparrow & & \uparrow f \\ Z & \xrightarrow[\beta]{} & Y \end{array}$$

$\underline{D}$ is then a partially ordered set, represented by the diagram

$$\begin{array}{ccc} & .h & \\ {}^{(\alpha,1)}\swarrow & & \searrow^{(\beta,1)} \\ f. & & .g \end{array}$$

Let $\underline{D}_o$ be the subcategory consisting of the single object f. The by definition (see [La 2] 4.2), $\mathrm{Def}(\underline{D}/\underline{D}_o)$ is the functor Def_1 define above.

The cohomology groups $A^i_{\underline{D}_o}(\underline{D},O_{\underline{D}})$ refered to above are defined as follows:

Given any category of morphisms of schemes $\underline{D}$ we may associate to it category of morphisms of affine schemes $\underline{d}$, simply by looking at the affine open subschemes of the schemes involved and all morphisms amon them induced by the objects of $\underline{D}$. Now given such a category of morphisms of affine schemes, or equivalently of morphisms of k-algebras we may define a new category $\mathrm{Mor}\ \underline{d}$ (see [La2] 1.1) the objects of which are the morphisms of $\underline{d}$, i.e. the commutative diagrams

$$\begin{array}{ccc} \mathrm{Spec}(A) & \xleftarrow[\alpha]{} & \mathrm{Spec}(A') \\ \mu\uparrow & & \uparrow\mu' \\ \mathrm{Spec}(B) & \xleftarrow[\beta]{} & \mathrm{Spec}(B') \end{array}$$

where μ and μ' are objects of $\underline{d}$ and (α,β) is a morphism $\mu' \to$
Given two such objects (α,β) and (α_1,β_1) a morphism $(\alpha,\beta)\to(\alpha_1,\beta$
in $\mathrm{Mor}\ \underline{d}$ is a pair of morphisms $((\varphi,\psi),(\varphi_1,\psi_1))$ of $\underline{d}$ making the following diagram commutative:

$$\begin{array}{ccccccc} \mathrm{Spec}(A_1) & \xleftarrow[\varphi_1]{} & \mathrm{Spec}(A) & \xleftarrow[\alpha]{} & \mathrm{Spec}(A') & \xleftarrow[\varphi]{} & \mathrm{Spec}(A'_1) \\ \uparrow\mu_1 & & \uparrow\mu & & \uparrow\mu' & & \uparrow\mu'_1 \\ \mathrm{Spec}(B_1) & \xleftarrow[\psi_1]{} & \mathrm{Spec}(B) & \xleftarrow[\beta]{} & \mathrm{Spec}(B') & \xleftarrow[\psi]{} & \mathrm{Spec}(B'_1) \end{array}$$

and such that $\varphi\alpha\varphi_1 = \alpha_1$, $\psi\beta\psi_1 = \beta_1$.

Given an object (α,β) of $\mathrm{Mor}\ \underline{d}$, consider the category A-free/B morphisms of A-algebras, $\delta : F \to B$, where F is a polynomial algeb over A, and consider the functor

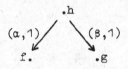

$$\mathrm{Der}_A(-,B') : (A\text{-}\underline{\mathrm{free}}/B)^{o} \longrightarrow \underline{\mathrm{Ab}}$$

defined by $\mathrm{Der}_A(F \to B, B') = \mathrm{Der}_A(\delta, B') = \mathrm{Der}_A(F, B')$ where B' is considered an F-module via the morphisms δ and $B \overset{\beta^*}{\to} B'$. The standard resolving complex for the projective limit functor, $C^{\cdot}(-,-)$, provides a complex of abelian groups (k-vectorspaces)

$$C^{\cdot}(\alpha,\beta) = C^{\cdot}((A\text{-}\underline{\mathrm{free}}/B)^{o}, \mathrm{Der}_A(-,B'))$$

and one checks that the map $(\alpha,\beta) \to C^{\cdot}(\alpha,\beta)$ is a functor

$$C^{\cdot}(O_{\underline{d}}) : \mathrm{Mor}\ \underline{d} \longrightarrow \underline{\text{complexes of k-vectorspaces}}$$

The cohomology of the total complex associated to the double complex

$$K^{\cdot\cdot}_{\underline{d}} = C^{\cdot}(\mathrm{Mor}\,\underline{d}, C^{\cdot}(O_{\underline{d}}))$$

is, by definition, the cohomology $A^{\cdot}(\underline{D}, O_{\underline{D}}) = A^{\cdot}(\underline{d}, O_{\underline{d}})$ of $\underline{D}$.

If $\underline{D}_o$ is a subcategory of $\underline{D}$, the corresponding category of morphisms of affine schemes $\underline{d}_o$ is a subcategory of $\underline{d}$. There is an obvious surjection of double complexes

$$K^{\cdot\cdot}_{\underline{d}} \longrightarrow K^{\cdot\cdot}_{\underline{d}_o}$$

the kernel of which is denoted by $K^{\cdot\cdot}_{\underline{d}/\underline{d}_o}$.

The cohomology we want, the algebraic cohomology of the pair $(\underline{D}, \underline{D}_o)$, is by definition the cohomology of this double complex, i.e.

$$A^{\cdot}_{\underline{D}_o}(\underline{D}, O_{\underline{D}}) = H^{\cdot}(K^{\cdot\cdot}_{\underline{d}/\underline{d}_o}).$$

To compute this cohomology we need the following lemma:

<u>Lemma</u>. Assume $\underline{D}$ as a category is an ordered set, such that for any morphism

$$(\alpha,\beta) : \mu \to \mu_1 : \begin{array}{ccc} X & \xrightarrow{\alpha} & X_1 \\ \mu\downarrow & & \downarrow\mu' \\ Y & \xrightarrow{\beta} & Y_1 \end{array}$$

of $\underline{D}$ the morphism β is affine. Then there is a spectral sequence given by the term

$$E_2^{p,q} = \varprojlim_{\mathrm{Mor}\,\underline{D}/\mathrm{Mor}\,\underline{D}_o}{}^{(p)} \underline{A}^q = H^p_{\underline{D}_o}(\underline{D},\underline{A}^q)$$

where the projective system $\underline{A}^q$ on Mor $\underline{D}$ is defined by

$$\underline{A}^q(\alpha,\beta) = A^q(\mu',R^{\cdot}\alpha_* O_X)$$

Proof. Since $\underline{D}$ is an ordered set, Mor $\underline{D}$ is an ordered set. Moreover $\underline{d}$ and Mor $\underline{d}$ are ordered sets. Given an object of Mor $\underline{D}$, say (α,β), let $\varkappa(\alpha,\beta)$ be the subset of Mor $\underline{d}$ consisting of the diagrams of the form

$$\begin{array}{ccc} \mathrm{Spec}(A) & \xrightarrow[\alpha_1]{} & \mathrm{Spec}(A_1) \\ \mu'\downarrow & & \downarrow\mu_1' \\ \mathrm{Spec}(B) & \xrightarrow[\beta_1]{} & \mathrm{Spec}(B_1) \end{array}$$

with Spec(A) open in X, Spec(B) open in Y, Spec(A') open in X and Spec(B') open in Y_1, μ', μ_1', α_1 and β_1 being the restrictions of μ, μ_1, α and β respectively.

Put: $\hat{\varkappa}(\alpha,\beta) = \{(\alpha_2,\beta_2) \in \mathrm{Mor}\,\underline{d} \mid \exists(\alpha_1,\beta_1) \in (\alpha,\beta)$ and $\exists$ morphism $(\alpha_1,\beta_1) \to (\alpha_2,\beta_2)$ in Mor $\underline{d}\}$

The conditions of the lemma imply that $\varkappa(\alpha,\beta)$ satisfies the conditions (i), (ii) and (iii) of ([La 1],(1.2.4)).

Therefore

$$\varprojlim_{\hat{\varkappa}(\alpha,\beta)}{}^{(q)} \simeq \varprojlim_{\varkappa(\alpha,\beta)}{}^{(q)} \qquad \text{for all } q \geq 0 .$$

The spectral sequence associated to the $\varkappa$-functor (see [La 1] (1.3.1))

$$\hat{\varkappa} : \mathrm{Mor}\,\underline{D} \to P\ \mathrm{Mor}\,\underline{d}$$

is given by the term

$$E_2^{p,q} = \varprojlim_{\underline{D}}{}^{(p)} \varprojlim_{\varkappa}{}^{(q)}$$

and converges to $\varprojlim_{\mathrm{Mor}\,\underline{d}}{}^{(\cdot)}$. From this the conclusion of the lemma follows immediately. Q.E.D.

For lack of a better place, we shall put here a lemma which we need in a joint paper with K. Lønsted published in this volume.

Lemma. Let $\pi : X \to S$ be a morphism of k-schemes, and let G be a group of automorphisms of X as S-scheme. Suppose there exists an open affine covering of X invariant under G, and consider the category $\underline{D}$ of morphisms of schemes containing one object, π, and for which the morphisms are the elements of G. Then there is a spectral sequence given by the term

$$E_2^{p,q} = H^p(G, A^q(\pi, O_X))$$

converging to $A^{\cdot}(\underline{D}, O_{\underline{D}})$.

Proof. We may assume S affine. Let $\underline{d}$ be the category of restrictions $\pi/U : U \to S$ where U is invariant under G, open and affine, the morphisms being compositions of elements of G and gluing morphisms.

It is easy to see that

$$\mathrm{Mor}\ \underline{d} \simeq \mathrm{Mor}\ \underline{G} \times \mathrm{Mor}\ \underline{d}'_{\pi}$$

where $\underline{G}$ is the category with one object and morphisms G, and $\underline{d}'_{\pi}$ is the category of G-invariant affine pieces of π, i.e. morphisms of the form $\pi/U : U \to S$ above, (see [La 2], (3.1.3)).

This implies the existence of a spectral sequence with $E_2^{p,q} = \varprojlim^{(p)}_{\mathrm{Mor}\ \underline{G}} \varprojlim^{(q)}_{\mathrm{Mor}\ \underline{d}'_{\pi}}$ converging to $\varprojlim^{(\cdot)}_{\mathrm{Mor}\ \underline{d}}$. Since $\varprojlim^{(\cdot)}_{\mathrm{Mor}\ \underline{G}} \simeq \varprojlim^{(\cdot)}_{\underline{G}} \simeq H^{(\cdot)}(G,-)$ the result now follows from the definitions. Q.E.D.

Let's apply the first lemma to our special situation. The category $\underline{D}$ is an ordered set, and it is easily seen that $\mathrm{Mor}\ \underline{D}$ is the ordered set corresponding to the diagram

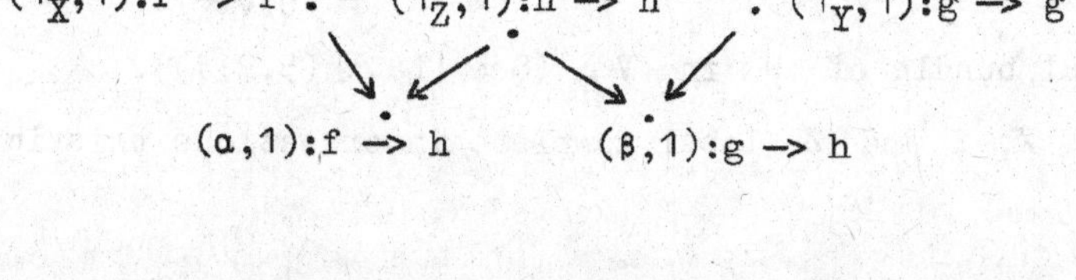

with $\text{Mor}\,\underline{D}_o$ reduced to the object $(1_X,1) : f \to f$.

Since all morphisms of schemes occuring in $\underline{D}$ are closed imbeddings, in particular affine, we obtain

$$\underline{A}^q(1_X,1) = A^q(f,O_X),\ \underline{A}^q(1_Z,1) = A^q(h,O_Z),\ \underline{A}^q(1_Y,1) = A^q(g,O_Y)$$

$$A^q(\alpha,1) = A^q(f,\alpha_* O_Z) \qquad \underline{A}^q(\beta,1) = A^q(g,\beta_* O_Z)$$

Thus the cohomology $A_{\underline{D}_o}(\underline{D},O_{\underline{D}})$ is the abutment of a spectral sequenc given by the term:

$$E_2^{p,q} = \varprojlim_{\text{Mor}\,\underline{D}/\text{Mor}\,\underline{D}_o}{}^{(p)} \underline{A}^q$$

$$= \varprojlim_{\text{Mor}\,\underline{D}/\text{Mor}\,\underline{D}_o}{}^{(p)} \left\{ \begin{array}{ccccc} A^q(f,O_X) & & A^q(h,O_Z) & & A^q(g,O_Y) \\ & k^q \searrow \ \swarrow & & 1^q \searrow m^q \ \swarrow n^q & \\ & A^q(f,\alpha_* O_Z) & & A^q(g,\beta_* O_Y) & \end{array} \right\}$$

$$= \begin{cases} \ker(1^q, m^q - n^q) & \text{for } p = 0 \\ \text{coker}(1^q, m^q - n^q) & \text{for } p = 1 \\ 0 & \text{for } p \geq 2 \end{cases}$$

(see [La 1] (1.2)).

Given any closed imbedding $i : U \to V$ and any O_U-Module F, the alg bra cohomology $A^\cdot(i,F)$ is the abutment of a spectral sequence given by the term:

$$E_2^{p,q} = H^p(V,\underline{A}^q(F))$$

where $\underline{A}^q(F)$ is the O_V-Module defined by

$$\underline{A}^q(F)(\text{Spec}(B)) = A^q(A,B;M)$$

whenever $\text{Spec}(B) \to \text{Spec}(A)$ is an affine piece of i and $F|\text{Spec}(B)$

Suppose U is a local complete intersection of V, then $\underline{A}^q(F) = 0$ for $q \neq 1$ and any F. Moreover $\underline{A}^1(F) \simeq N_{U/V} \otimes_{O_U} F$ where $N_{U/V}$ is the normal bundle of U in V. (See [La 2] (5.2.1)).

Assuming X, Y and Z local complete intersections this implies the following isomorphisms:

$$A^1_{\underline{D}_o}(\underline{D},O_{\underline{D}}) \simeq E_2^{o,1} \simeq \ker\{(l^1,m^1-n^1): H^o(Z,N_Z)\oplus H^o(Y,N_Y) \longrightarrow H^o(X,N_X\otimes O_Z)\oplus H^o(Y,N_Y\otimes O_Z)\}$$

$$A^2_{\underline{D}_o}(\underline{D},O_{\underline{D}}) \simeq E_2^{1,1}\oplus E_2^{o,2} \simeq \operatorname{coker}(l^1,m^1-n^1)\oplus\ker(l^2,m^2-n^2)\,.$$

We conclude that there is an isomorphism

$$\hat{O}_{I,y} \simeq \operatorname{Sym}_k(A^1_{\underline{D}_o}(\underline{D},O_{\underline{D}})^*)\ \hat{\underset{\operatorname{Sym}_k(A^2_{\underline{D}_o}(\underline{D},O_{\underline{D}})^*)^\wedge}{\otimes}}\ k$$

where the right hand side is defined by the above formulas and by the total obstruction morphism

$$o : \operatorname{Sym}_k(A^2_{\underline{D}_o}(\underline{D},O_{\underline{D}})^*)^\wedge \longrightarrow \operatorname{Sym}_k(A^1_{\underline{D}_o}(\underline{D},O_{\underline{D}})^*)^\wedge$$

of which we know very little, see [La 3].

Now consider the diagram (**) and the corresponding categories $\underline{D}'$ and $\underline{D}'_o$. In exactly the same way as above we compute the algebra cohomology $A^{\cdot}_{\underline{D}_o}(\underline{D}';O_{\underline{D}'})$,

$$A^1_{\underline{D}'_o}(\underline{D}';O_{\underline{D}'}) \simeq \ker\{(l^1,m^1-n^1): H^o(Z',N_{Z'/L})\oplus H^o(Y,N_{Y/L}) \longrightarrow H^o(X',N_{X'/L})\otimes O_{Z'})\oplus H^o(Y,N_{Y/L}\otimes O_{Z'})\}$$

$$A^2_{\underline{D}'_o}(\underline{D}';O_{\underline{D}'}) \simeq \operatorname{coker}(l^1,m^1-n^1)\oplus\ker(l^2,m^2-n^2)\,.$$

For the diagram (***) we find in particular:

$$A^1_{\underline{D}''_o}(\underline{D}'';O_{\underline{D}''}) \simeq H^o(L,N_{L/\mathbb{P}^N})$$

$$A^2_{\underline{D}''_o}(\underline{D}'';O_{\underline{D}''}) = 0\,.$$

In this case one should consider the categories $\underline{D}''$ and $\underline{D}''_o$ corresponding to the diagram

$$\begin{array}{ccccc} X=\emptyset & \longrightarrow & \mathbb{P}^N & \longleftarrow & L=Y \\ & \nwarrow & \uparrow & \nearrow & \\ & & \emptyset = Z & & \end{array}$$

As above we have isomorphisms of complete local k-algebras:

$$\hat{O}_{g^{-1}(L),y} \simeq \mathrm{Sym}_k(A^1_{\underline{D}_o}(\underline{D}';O_{\underline{D}'})^*)\hat{\otimes}_{\mathrm{Sym}_k(A^2_{\underline{D}'_o}(\underline{D}';O_{\underline{D}'})^*)^{\wedge}} k$$

$$\hat{O}_{G,L} \simeq \mathrm{Sym}_k(H^o(L,N_{L/\mathbb{P}^N})^*)^{\wedge}$$

Note also that the morphisms $\mathbb{j}^*$ and $\mathbb{g}^*$ correspond at the tangent space level, to the canonical morphisms

$$\mathbb{j}^t : A^1_{\underline{D}'_o}(\underline{D}';O_{\underline{D}'}) \longrightarrow A^1_{\underline{D}_o}(\underline{D},O_{\underline{D}})$$

and

$$\mathbb{g}^t : A^1_{\underline{D}_o}(\underline{D};O_{\underline{D}}) \longrightarrow A^1_{\underline{D}''_o}(\underline{D}'';O_{\underline{D}''}) = H^o(L,N_{L/\mathbb{P}^N}),$$

induced by the morphisms

$$N_{Z/L} \longrightarrow N_{Z/\mathbb{P}^N}$$

$$N_{Y/L} \longrightarrow N_{Y/\mathbb{P}^N}$$

$$N_{X'/L} \longrightarrow N_{X/\mathbb{P}^N}$$

and the morphism

$$H^o(Y,N_{Y/\mathbb{P}^N}) \longrightarrow H^o(L,N_{L/\mathbb{P}^N})$$

induced by the inclusion $C(1)_y \subseteq C_y$ (see § 1), where Y corresponds to the point $y \in I_o$. We know that the cone of $\tilde{I}_o$ is a deformation of the cone of Y to the pointed scheme $\{y\} \to \tilde{I}_o$. Therefore the formalization of the cone of $\tilde{I}_o$ at the point $y \in I_o$ is the formal cone of the universal formal family

$$\begin{array}{ccc} \tilde{I}_o^{\wedge} & \hookrightarrow & \mathbb{P}^N \otimes \hat{O}_{I_o,y} \\ \downarrow & & \\ \mathrm{Spf}(\hat{O}_{I_o,y}) & & \end{array}$$

The homogenous ideal of this imbedding, $C_y^{\wedge}$ is therefore the formalization at y of C.

Now $\hat{O}_{I_o,y}[X_0,\ldots,X_N]/C_y^{\hat{}}$ being a graded deformation of the graded k-algebra $k[X_0,\ldots,X_N]/C_y$ to a complete local k-algebra (with residue field k), there corresponds a unique morphism

$$\rho : H_Y \longrightarrow \hat{O}_{I_o,y}$$

where H_Y is the formal moduli of $k[X_0,\ldots,X_N]/C_y$ in the graded sense (see [K]). The tangent space of H_Y, $(\underline{m}_H/\underline{m}_H^2)^*$ is isomorphic to

$$A^1_{grad.}(k[X_0,\ldots,X_N],k[X_0,\ldots,X_N]/C_y\,;\,k[X_0,\ldots,X_N]/C_y)$$
$$= \mathrm{Hom}^{grad.}_{k[X_0,\ldots,X_N]}(C_y/C_y^2,k[X_0,\ldots,X_N]/C_y) \simeq H^o(Y,N_{Y/\mathbb{P}^N})$$

(remember Y is a complete intersection). The morphism ρ' is, at the tangent level, the canonical morphism $\rho^t : A^1_{\underline{D}_o}(\underline{D},O_{\underline{D}}) \to H^o(Y,N_{Y/\mathbb{P}^N})$ defined by restriction.

The formal universal family

$$H_Y[X_0,\ldots,X_N]/\bar{C}_y$$

is given in the following way. Let C_y be generated by the regular sequence of homogenous polynomials $(F_1,\ldots,F_s)$, and let $\{Z_1,\ldots,Z_t\}$ be a basis of $H^o(Y,N_{Y/\mathbb{P}^N}) = \mathrm{Hom}^{grad}(C_y,k[X_0,\ldots,X_N]/C_y)$. Then

$$H_Y = k[[Z_1^*,\ldots,Z_t^*]]$$

and $\bar{C}_y$ is generated by

$$\bar{F}_i = F_i + \sum_{j=1}^{t} Z_j'(F_i)Z_j \qquad i=1,\ldots,s$$

where $Z_j'(F_i)$ is any representative in $k[X_0,\ldots,X_N]$ of $Z_j(F_i) \in k[X_0,\ldots,X_N]/C_y$.

Let $\{y_1,\ldots,y_p\}$ be a basis of $A^1_{\underline{D}_o}(\underline{D},O_{\underline{D}})$, then $\{y_1^*,\ldots,y_p^*\}$ generates the maximal ideal of $\hat{O}_{I_o,y}$. Since ρ maps $\bar{C}_y$ onto $C_y^{\hat{}}$ we find that modulo $(y_1^*,\ldots,y_p^*)^2$, $C_y^{\hat{}}$ is generated by the forms

$$F_i^{\hat{}} = F_i + \sum_{k=1}^{p} y_k(F_i)y_k^* \qquad i=1,\ldots,s,$$

where we by $y_k(F_i)$ actually mean

$$\rho^t(y_k)'(F_i).$$

Finally, let's remark that what we have shown up to now proves that I has imbeddingdimension $\dim_k A^1_{\underline{D}_o}(\underline{D},O_{\underline{D}})$ at y, that the dimension of I at y is bounded below by $\dim_k A^1_{\underline{D}_o}(\underline{D},O_{\underline{D}}) - \dim_k A^2_{\underline{D}_o}(\underline{D},O_{\underline{D}})$, and that I is nonsingular at y if $A^2_{\underline{D}_o}(\underline{D},O_{\underline{D}}) = 0$.

To go further we have to put more conditions on $\mathbb{h}$, $\mathbb{h}_o$ and, in particular on Y and Z.

§ 3. The case of r-secants.

Let $\mathbb{h}$ be $\mathrm{Grass}(2,N+1) \subseteq \mathrm{Hilb}_{\mathbb{P}^N}$ and let $\mathbb{h}_o$ be $\mathrm{Hilb}^r_{\mathbb{P}^N} \subseteq \mathrm{Hilb}_{\mathbb{P}^N}$ the subscheme of $\mathrm{Hilb}_{\mathbb{P}^N}$ parametrizing the finite closed subschemes of length r of $\mathbb{P}^N$, and consider the universal diagram

$$\begin{array}{ccccc} \tilde{I} & \xrightarrow[\tilde{g}]{} & \mathbb{P}^N \times I & \xrightarrow[pr_1]{} & \mathbb{P}^N \\ {\scriptstyle\pi}\downarrow & & & & \\ I & & & & \end{array}$$

Put $\varphi_r = \tilde{g}\, pr_1$, then we make the following

Definition. $I = \mathrm{Sec}_r(X)$ is the r-secant moduli scheme, $\pi : \tilde{I} \to \mathrm{Sec}_r(X)$ is the r-secant bundle, $\varphi_r : \tilde{I} \to \mathbb{P}^N$ is the r-secant morphism and $\mathrm{im}\, \varphi_r = S_r(X)$ is the r-secant scheme.

There are several invariants associated to this situation. The first one is

$$d_r = \dim \mathrm{Sec}_r(X).$$

Next, one would like to compute

$$s_r = \dim S_r(X).$$

In case $s_r = d_r + 1 = N$, there is an open subset of $\mathbb{P}^N$ through every point of which there passes the same number of r-secants. This number δ_r is simply the degree of φ_r, and has been computed by several authors, see [H],[P],[L].

We shall concentrate on the invariants d_r and s_r in some nice cases.

Let $L \in \mathrm{Sec}_r(X)$ be any r-secant, we know that

$$\hat{O}_{\mathrm{Sec}_r(X),L} \simeq T^1 \hat{\otimes}_{T^2} k$$

where $T^i = \mathrm{Sym}_k(A^i_{\underline{D}_o}(\underline{D},O_{\underline{D}})^*)^\wedge$ and

$$A^1_{\underline{D}_o}(\underline{D},O_{\underline{D}}) = \{(\underline{v},\underline{w}) \in H^o(Z,N_Z) \oplus H^o(L,N_L) \mid l(\underline{v}) = 0, m(\underline{v}) = n(\underline{w})\}$$

$$A^2_{\underline{D}_o}(\underline{D},O_{\underline{D}}) = H^o(X,N_X \underset{O_X}{\otimes} O_Z) \oplus H^o(L,N_L \underset{O_L}{\otimes} O_Z)/\mathrm{im}(l,m-n).$$

The last formulas follow from § 2 since in this case we have $H^1(L,N_L) = 0, H^1(Z,N_Z) = 0$.

To compute the A^i's we will have to make explicite the diagram

$$\begin{array}{ccccc} H^o(X,N_X) & & H^o(Z,N_Z) & & H^o(L,N_L) \\ {\scriptstyle k}\searrow & \swarrow{\scriptstyle l} & & {\scriptstyle m}\searrow & \swarrow{\scriptstyle n} \\ & H^o(X,N_X \otimes O_Z) & & H^o(L,N_L \otimes O_Z) & . \end{array}$$

It is easily seen that

$$N_L = \underbrace{O_L(1) \oplus \ldots \oplus O_L(1)}_{N-1}$$

Suppose $Z = X \cap L = \{P_1,\ldots,P_r\}$ is nonsingular, and that the P_i's are nonsingular k-rational points of X. Then

$$H^o(L,N_L) \simeq k^{(N-1)}$$

$$H^o(Z,N_Z) = \coprod_{i=1}^{r} H^o(P_i,N_{P_i}) \simeq k^{rN}$$

$$H^o(L,N_L \otimes O_Z) = \coprod_{i=1}^{r} H^o(L,N_L \otimes O_{P_i}) \simeq k^{r(N-1)}$$

$$H^o(X,N_X \otimes O_Z) = \coprod_{i=1}^{r} H^o(X,N_X \otimes O_{P_i}) \simeq k^{r(N-d)}$$

where $d = \dim X$, assuming X equidimensional.

Counting dimensions we obtain

$$\dim_k A^1_{\underline{D}_o}(\underline{D},O_{\underline{D}}) - \dim_k A^2_{\underline{D}_o}(\underline{D},O_{\underline{D}}) = 2(N-1) - r(N-d-1)$$

Thus

$$\dim \mathrm{Sec}_r(X) \geq 2(N-1) - r(N-d-1).$$

Since $\dim_k A^1_{\underline{D}_o}(\underline{D},O_{\underline{D}}) = 2(N-1) - r(N-d-1)$ implies $A^2_{\underline{D}_o}(\underline{D},O_{\underline{D}}) = 0$ and vice versa, we observe that if L is a nonsingular point of $\mathrm{Sec}_r(X)$ the equality holds above.

The morphisms l and m are simply the (obvious) projection of a vector at P_i onto the normal space of X respectively L at P_i. If $(\underline{v},\underline{w})$ is an element of $A^1_{\underline{D}_o}(\underline{D},O_{\underline{D}})$, then $\underline{v} = \{\underline{v}_i\}_{i=1}^r$, $\underline{v}_i \in H^o(P_i,N_{P_i})$ and $l(\underline{v}) = 0$ means that the vectors $\underline{v}_i$ are tangent vectors of X at P_i.

Suppose $\underline{v}_i \in H^o(P_i,N_{P_i})$ is tangent to X at P_i, then $m(\underline{v}_i)$, being the projection of $\underline{v}_i$ onto the normal space of L at P_i, is nonzero provided $\underline{v}_i \neq 0$. In fact, if $m(\underline{v}_i) = 0$, then $\underline{v}_i$ must be tangent to L at P_i and therefore L must be tangent to X at P_i which is impossible since $X \cap L$ is supposed to be nonsingular.

Moreover, given $\underline{w} \in H^o(L,N_L)$, to say that there exists a $\underline{v} \in H^o(Z,N_Z)$ with $m(\underline{v}) = n(\underline{w})$ and $l(\underline{v}) = 0$ is equivalent to saying that the ith component $n(w)_i$ of $n(\underline{w})$ sits in the linear subspace $\underline{M}(P_i)$ of the tangentspace $\mathbf{T}(P_i)$ of $\mathbb{P}^N$ at P_i, generated by the tangent of L at P_i and the tangentspace of X at P_i, $i = 1,\ldots,r$.

Pick an innerproduct on $\mathbf{T}(P_i)$ and let $\underline{N}(P_i)$ be the normal subspace $\underline{M}(P_i)^{\perp}$, then we have proved the following,

$$\begin{aligned} A^1_{\underline{D}_o}(\underline{D},O_{\underline{D}}) &= \{(\underline{v},\underline{w}) \in H^o(Z,N_Z) \oplus H^o(L,N_L) \mid l(\underline{v}) = 0, m(\underline{v}) = n(\underline{w})\} \\ &= \{\underline{w} \in H^o(L,N_L) \mid n(\underline{w})_i \in \underline{M}(P_i),\ i = 1,\ldots,r\} \\ &= \{\underline{w} \in H^o(L,N_L) \mid n(\underline{w})_i \perp \underline{N}(P_i),\ i = 1,\ldots,r\} \end{aligned}$$

Let $(\alpha_o(i),\alpha_1(i),0,\ldots,0)$ be homogenous coordinates of P_i. We may suppose all $\alpha_o(i) \neq 0$.

Consider

$$H^o(L,N_L) = H^o(L,O_L(1)^{N-1})$$

$$= \mathrm{Hom}((X_2,\dots,X_N)/X_2,\dots,X_N)^2, k[X_o,X_1])_o$$

and denote by $\{\underline{w}_{jl}\}_{j=2,\dots,N,\,l=o,1}$ the basis for $H^o(L,N_L)$ given by

$$\underline{w}_{jl}(X_k) = \begin{cases} 0 & k \neq j \\ X_o & k = j,\ l = 0\ , \\ X_1 & k = j,\ l = 1 \end{cases} \qquad k = 2,\dots,N .$$

Let

$$\{\underline{n}_{X_j}(P_i)\}_{j=1,\dots,N}$$

be the basis of the normal space of L at P_i such that:

$$\underline{n}_{X_j}(P_i) \in H^o(L,N_L \otimes O_{P_i})$$

$$= \mathrm{Hom}({}^{(X_2,\dots,X_N)}/{}_{(X_2,\dots,X_N)^2},\ {}^{k[X_o,X_1]}/{}_{(\alpha_1(i)X_o - \alpha_o(i)X_1)})_o$$

$$= \mathrm{Hom}({}^{(X_2,\dots,X_N)}/{}_{(X_2,\dots,X_N)^2}, k[X_o]))_o \quad ,$$

is defined by

$$\underline{n}_{X_j}(X_l) = \underline{n}_{X_j}(P_i)(X_l) = \begin{cases} 0 & l \neq j \\ X_o & l = j \end{cases}$$

An easy computation shows that the morphism

$$n : H^o(L,N_L) \longrightarrow H^o(L,N_L \otimes O_Z) = \coprod_{i=1}^{r} H^o(L,N_L \otimes O_{P_i})$$

is given by: $n = (n_i)_{i=1,\dots,r}$.

$$n_i(\underline{w}_{jl}) = \begin{cases} \underline{n}_{X_j}(P_i) & l = 0 \\ \dfrac{\alpha_1(i)}{\alpha_o(i)}\, \underline{n}_{X_j}(P_i) & l = 1 \end{cases}$$

Now, with these notations we observe that

$$A^1_{\underline{d}_o}(\underline{d},O_{\underline{d}}) \simeq \{\, \Sigma u_{jl}\underline{w}_{jl} \in H^o(L,N_L) \mid (\sum_{j=2}^{N} (\alpha_o(i)u_{jo} + \alpha_1(i)u_{j1})\underline{n}_{X_j}) \cdot \underline{n}(i) = 0$$

$$\text{for all } \underline{n}(i) \in \underline{N}(P_i) \text{ and all } i \,\} .$$

The conditions on u_{jl} $j=2,\dots,n$, $l=0,1$ expressed by the equations

(1) $$\Big(\sum_{j=2}^{N}(\alpha_o(i)u_{jo}+\alpha_1(i)u_{j1})\underline{n}_{X_j}\Big)\cdot\underline{n}(i) = 0$$

are equivalent to the following: Put

$$\begin{pmatrix} n(1)_{1,2} & n(1)_{1,3} & \cdots & n(1)_{1,N} \\ \vdots & \vdots & & \vdots \\ n(1)_{N-d-1,2} & n(1)_{n-d-1,3} & \cdots & n(1)_{n-d-1,N} \\ \hline n(2)_{1,2} & n(2)_{1,3} & \cdots & n(2)_{1,N} \\ \vdots & & & \\ n(2)_{N-d-1,2} & n(2)_{N-d-1,3} & \cdots & n(2)_{N-d-1,n} \\ \hline & \vdots & & \vdots \\ \hline n(r)_{1,2} & n(r)_{1,3} & \cdots & n(r)_{1,N} \\ \vdots & \vdots & & \vdots \\ n(r)_{N-d-1,2} & n(r)_{N-d-1,3} & \cdots & n(r)_{N-d-1,n} \end{pmatrix} = \begin{pmatrix} \underline{N}_1 \\ \hline \underline{N}_2 \\ \hline \vdots \\ \hline \underline{N}_r \end{pmatrix}$$

where

$$\underline{n}(i)_k = (n(i)_{k,2}, n(i)_{k,3}, \dots, n(i)_{k,N}) \in \underline{N}(P_i),$$

$k=1,\dots,N-d-1$ is a basis for $\underline{N}(P_i)$.

Remember that $\underline{N}(P_i) \subseteq [\underline{n}_{X_2}, \underline{n}_{X_3}, \dots, \underline{n}_{X_N}]$.

Then (1) is equivalent to the following system of linear equations

(2) $$\begin{pmatrix} \alpha_o(1)\underline{N}_1 & \alpha_1(1)\underline{N}_1 \\ \hline \alpha_o(2)\underline{N}_2 & \alpha_1(2)\underline{N}_2 \\ \hline \vdots & \vdots \\ \hline \alpha_o(r)\underline{N}_r & \alpha_1(r)\underline{N}_r \end{pmatrix} \begin{pmatrix} u_{2,o} \\ \vdots \\ u_{N,o} \\ \hline u_{2,1} \\ \vdots \\ u_{\underline{N},1} \end{pmatrix} = 0$$

Put $\alpha_i = \alpha_1(i)/\alpha_o(i)$ and consider the linear subspace

$$Q = \{(\underline{n}_i)_{i=1}^r \in \coprod_{i=1}^r \underline{N}(P_i) \mid \sum_{i=1}^r \underline{n}_i = \sum_{i=1}^r \alpha_i \underline{n}_i = 0\}$$

Obviously the rank of the matrix of (2) is $r(N-d-1) - \dim_k Q$. Put $\dim_k Q = \varepsilon$, then

$$\dim_k A^1_{\underline{D}_o}(\underline{D}, O_{\underline{D}}) = 2(N-1) - r(N-d-1) + \varepsilon$$

$$\dim_k A^2_{\underline{D}_o}(\underline{D}, O_{\underline{D}}) = \varepsilon .$$

We shall use this to prove the following well known result, see [M], [Ab], [An], [Sa].

<u>The trisecant lemma</u>. Let X be an irreducible reduced curve of $\mathbb{P}^3$. Assume $\operatorname{char} k = 0$ (or big enough). Then $\operatorname{Sec}_3(X) \neq \emptyset$ implies $d_3 = \dim \operatorname{Sec}_3(X) = 1$ unless X is plane of degree 3, in which case $d_3 = 2$ and $\operatorname{Sec}_3(X)$ is nonsingular.

<u>Proof</u>. Let $L \in \operatorname{Sec}_3(X)$ and suppose $\dim \operatorname{Sec}_3(X) \geq 2$ at L. Then *)

$$\dim_k A^1_{\underline{D}_o}(\underline{D}, O_{\underline{D}}) = 4 - 3 + \varepsilon \geq 2$$

implying $\varepsilon \geq 1$. This, however, implies that all $\underline{N}(P_i)$ $i = 1,2,3$ coincide as subspaces of $[\underline{n}_{X_2}, \dots, \underline{n}_{X_N}]$, and $\varepsilon = 1$. Therefore $\operatorname{Sec}_3(X)$ has dimension 2 and imbedding dimension 2 at L, thus $\hat{O}_{\operatorname{Sec}_3(X),L} \simeq T^1 = k[y_o^*, y_1^*]$, where $\{y_o, y_1\}$ is a basis of $A^1_{\underline{D}_o}(\underline{D}, O_{\underline{D}})$. Now y_j, $j = 0,1$ correspond to elements of $\operatorname{Hom}((x_2,x_3)/(x_2,x_3)^2, k[x_o,x_1])_o$. We may obviously assume $y_j(x_3) = x_j$, $j = 0,1$. Put $y_j(x_2) = l_j$, $j = 0,1$.
Then we have seen that the closed imbedding

$$\begin{array}{c} \tilde{I} \subseteq \mathbb{P}^3 \times \operatorname{Sec}_3(X) \\ \downarrow \\ \operatorname{Sec}_3(X) \end{array}$$

formally at $L \in \operatorname{Sec}_3(X)$ is defined by an ideal

*) Since the set of tangents to X has dimension 1, we may assume that L cuts X in 3 different points.

$$(x_2^\wedge, x_3^\wedge) \subseteq k[[y_o^\wedge, y_1^\wedge]][x_o, x_1, x_2, x_3]$$

where

$$x_2^\wedge \equiv x_2 + l_o y_o^* + l_1 y_1^* \pmod{(y_o^*, y_1^*)^2}$$

$$x_3^\wedge \equiv x_3 + x_o y_o^* + x_1 y_1^* \pmod{(y_o^*, y_1^*)^2}$$

Put $\tilde{Z} = \tilde{I}_{\mathbb{P}^3 \times \mathrm{Sec}_3(X)} \times (X \times \mathrm{Sec}_3(X))$. Then the closed imbedding

$$\begin{array}{l} \tilde{Z} \subseteq \mathbb{P}^3 \times \mathrm{Sec}_3(X) \\ \downarrow \\ \mathrm{Sec}_3(X) \end{array}$$

is defined formally at $L \in \mathrm{Sec}_3(X)$, by an ideal of the form

$$\prod_{i=1}^{r} (x_1 - \alpha_1^\wedge(i) x_o, x_2 - \alpha_2^\wedge(i) x_o, x_3 - \alpha_3^\wedge(i) x_o)$$

of $k[[y_o^* y_1^*]][x_o, x_1, x_2, x_3]$, where $\alpha_j^\wedge(i) \in k[[y_o^*, y_1^*]]$ and $\alpha_j^\wedge(i) \equiv \frac{\alpha_j(i)}{\alpha_o(i)} \pmod{(y_o^*, y_1^*)}$. (But this ideal is very far from the ideal defining the projecting cone of $\tilde{Z}^\wedge$.)

Since L intersect X transversally, we may assume that the plane $x_3 = 0$ also intersect X transversally in all P_i, $i = 1,2,3$. Moreover, since we have picked a coordinate system such that all P_i, $i = 1,2,3$ are contained in the affine piece of $\mathbb{P}^3$ defined by $x_o \neq 0$, we may as well dehomogenize at x_o.

Let $\mathfrak{a}$ be the ideal of X in $k[\frac{x_1}{x_o}, \frac{x_2}{x_o}, \frac{x_3}{x_o}] = k[x_1, x_2, x_3]$. Then the ideal $(\mathfrak{a} + (x_3)) \cdot k[x_1, x_2, x_3]_{P_i}$ is the maximal ideal $\underline{m}_{P_i}$ of $k[x_1, x_2, x_3]_{P_i}$ for $i = 1,2,3$.

Let $\underline{n} = n_2 \underline{n}_{x_2} + n_3 \underline{n}_{x_3}$ be any non-zero element of $\underline{N}(P_1) = \underline{N}(P_2) = \underline{N}(P_3)$. Then the plane $n_2 x_2 + n_3 x_3$ is tangent to X at P_i, $i = 1,2,3$. Therefore $n_2 x_2 + n_3 x_3 \in \mathfrak{a} + \underline{m}_{P_i}^2$, i.e. $n_2 x_2 + n_3 x_3 = a_i + R_i x_3^2$ where $a_i \in \mathfrak{a}$, $R_i \in k[x_1, x_2, x_3]_{P_i}$, $i = 1,2,3$.

Since $x_3 \equiv x_3^\wedge - (y_o^* + x_1 y_1^*) \pmod{(y_o^*, y_1^*)^2}$ we find $x_3^2 \equiv (y_o^* + x_1 y_1^*)^2 \pmod{(x_3^\wedge) + (y_o^*, y_1^*)^3}$.

On the other hand we know that

$$n_2x_2 + n_3x_3 = n_2\hat{x}_2 + n_3\hat{x}_3 + \sum_{p,q=o}^{1} l_{p,q}y_p^*y_q^* \pmod{(y_o^*,y_1^*)^3}$$

where since $\operatorname{char} k \neq 2$ we may assume $l_{p,q} = l_{q,p}$ are polynomials in x_1 of degree 1. Therefore we get the following congruence

$$(3) \qquad \sum_{p,q=o}^{1} l_{p,q}y_p^*y_q^* \equiv a_i + R_i(y_o^*+x_1y_1^*)^2 \pmod{(\hat{x}_2,\hat{x}_3)+(y_o^*,y_1^*)^3}$$

Since $\tilde{Z} \subseteq X \times \operatorname{Sec}_3(X)$ we have

$$a(\hat{\alpha}_1(i),\hat{\alpha}_2(i),\hat{\alpha}_3(i)) = 0 \quad \text{for } i = 1,2,3$$

and all $a(x_1,x_2,x_3) \in \mathfrak{a}$. Evaluating (3) at $(\hat{\alpha}_1(i),\hat{\alpha}_2(i),\hat{\alpha}_3(i))$ we obtain:

$$R_i(\alpha_1(i),0,0)(y_o^*+\alpha_1(i)y_1^*)^2 = \sum_{p,q} l_{p,q}(\alpha_1(i))y_p^*y_q^*$$

which is equivalent to

$$l_{p,q}(P_i) = x_1^{p+q}(P_i) \cdot R_i(P_i) \qquad i = 1,2,3,$$

or

$$l_{p,q}(P_i) = x_1^{p+q}(P_i) \cdot l_{o,o}(P_i) \qquad i = 1,2,3.$$

From this we deduce the following: The polynomials

$$l_{o,1} - x_1 l_{o,o} \qquad \text{and} \qquad l_{1,1} - x_1 l_{o,1}$$

both have 3 different roots $\alpha_1(i)$, $i = 1,2,3$. This implies $l_{p,q} = 0$ for all $p,q = 0,1$ and in particular $R_i(P_i) = 0$, $i = 1,2,3$.

Thus $n_2x_2 + n_3x_3 \equiv n_2\hat{x}_2 + n_2\hat{x}_3 + \sum_{o}^{1} l_{p,q}y_p^*y_q^*y_r^* \pmod{(y_o^*,y_1^*)^4}$ and $R_i \in \underline{m}_{P_i} = (\mathfrak{a} + (x_3)) \cdot k[x_1,x_2,x_3]_{P_i}$, which implies $R_i = a_i' + R_i'x_3$ with $a_i' \in \mathfrak{a} \cdot k[x_1,x_2,x_3]_{P_i}$, $R_i' \in k[x_1,x_2,x_3]$. From this we deduce

$$n_2x_2 + n_3x_3 = a_i + R_ix_3^2 = (a_i + a_i'x_3^2) + R_i'x_3^3.$$

Evaluating the congruence

$$\sum_{o}^{1} l_{p,q,r}y_p^*y_q^*y_r^* \equiv (a_i + a_i'x_3^2) + R_i'(y_o^*+x_1y_1^*)^3 \pmod{(\hat{x}_2,\hat{x}_3) + (y_o^*,y_1^*)^4}$$

at $(\hat{\alpha}_1(i),\hat{\alpha}_2(i),\hat{\alpha}_3(i))$ we find

$$l_{p,q,r}(P_i) = x_1^{p+q+r} R_i'(P_i) \qquad \text{for } i = 1,2,3,$$

provided we have arranged it such that $l_{p,q,r}$ is symmetric in p,q,r (Here is where the condition on the characteristic enters.)

From this follows as above $l_{p,q,r} = 0$ for all p,q,r and $R_i'(P_i) = ($ for $i = 1,2,3$.

Go on; repeating this process will eventually prove

$$n_2x_2 + n_3x_3 = n_2\hat{x_2} + n_3\hat{x_3} \qquad \text{i.e.}$$

$$n_2x_2 + n_3x_3 \in C_L^{\hat{}} .$$

But, by faithfully flatness, this implies

$$n_2x_2 + n_3x_3 \in \tilde{C}$$

proving that $n_2x_2 + n_3x_3$ is an equation for

$$S_3(X) \subseteq \mathbb{P}^3 .$$

Since $S_3(X)$ contains infinitely many points of X, we may conclude

$$X \subseteq S_3(X) \subset V(n_2x_2 + n_3x_3) ,$$

i.e. X is a plane curve. Since $\mathrm{Sec}_3(X) \neq \emptyset$, X has degree 3 and $S_3(X) = V(n_2x_2 + n_3x_3)$. The rest is obvious. Q.E.D.

Remark. We have seen that the proof of this lemma depends upon char k being big enough. One would of course like to know how big. See [Ab] [An] for further information on the char k = p case.

§ 4. A generalized trisecant lemma.

Let $X \subseteq \mathbb{P}^3$ be an irreducible reduced curve, and consider the closed subschemes of $\mathrm{Hilb}_{\mathbb{P}^3}$, $\mathbb{h}$, parametrizing the plane curves of degree n, and $\mathbb{h}_o = \mathrm{Hilb}^r_{\mathbb{P}^3}$.

Put $\mathbb{P}^3 = \mathrm{Proj}(k[x_o,x_1,x_2,x_3])$ and let $\mathfrak{a}$ be the homogenous ideal of $k[x_o,x_1,x_2,x_3]$ of forms vanishing on X. Let $Y \in \mathbb{h}$ and assume $Y = V(x_3,f)$ where $f \in k[x_o,x_1,x_2]$ is a form of degree n.

Suppose the intersection $Z = X \cap Y = \{P_1,\ldots,P_r\}$ is nonsingular and that P_i is a nonsingular retional point on X and on $X \cap V(X_3)$.

We may assume $Z \subseteq D(x_o) = \mathrm{Spec}(k[\frac{x_1}{x_o},\frac{x_2}{x_o},\frac{x_3}{x_o}]) = \mathrm{Spec}(k[x_1,x_2,x_3])$.

Denote by $\mathfrak{a}$ the ideal $\overline{\mathfrak{a}}$ dehomogenized at x_o.

Consider the corresponding diagram

$$\begin{array}{ccc} \tilde{I} & \longrightarrow & \mathbb{P}^3 \times I \\ {\scriptstyle\pi}\downarrow & & \\ I = I(X;h,h_o) & & \end{array}$$

Since Y is a complete intersection, the morphism

$$\mathfrak{g} : I_o \longrightarrow \mathrm{Grass}(3.4)$$

is defined in a neighbourhood of Y.

We shall study $\mathfrak{g}$ locally at the point Y of I_o.

We know that

$$\hat{O}_{I,Y} = T^1 \hat{\otimes}_{T^2} k$$

where $T^i = \mathrm{Sym}_k(A^i_{\underline{D}_o}(\underline{D},O_{\underline{D}})^*)\hat{}$. The first problem is therefore to compute the cohomology groups $A^i_{\underline{D}_o}(\underline{D},O_{\underline{D}})$. As above (see §3) we have to study the diagram

$$\begin{array}{ccccccc} H^o(X,N_X) & & & H^o(Z,N_Z) & & & H^o(Y,N_Y) \\ & {\scriptstyle k^o}\searrow & \swarrow{\scriptstyle l^o} & & \searrow{\scriptstyle m^o} & \swarrow{\scriptstyle n^o} & \\ & H^o(X,N_X\otimes O_X) & & & & H^o(Y,N_Y\otimes O_Z) & \end{array}$$

Here

$$H^o(Z,N_Z) = \coprod_{i=1}^{r} H^o(P_i,N_{P_i})$$

$$H^o(Y,N_Y) \simeq \mathrm{Hom}((f,x_3)/(f,x_3)^2, k[x_o,x_1,x_2,x_3,]/(f,x_3))_o$$

$$\simeq k[x_o,x_2,x_2]_{(1)} \oplus k[x_o,x_1,x_2]_{(n)}/(f)$$

$$H^o(Y,N_Y\otimes O_Z) = \coprod_{i=1}^{r} H^o(Y,N_Y\otimes O_{P_i})$$

$$H^o(Y,N_Y\otimes O_{P_i}) \simeq \mathrm{Hom}((f,x_3)/(f,x_3)^2, k[x_o,x_1,x_2,x_3]/(t_1(i),t_2(i),t_3(i)))_o$$

$$\simeq \mathrm{Hom}((f,x_3)/(f,x_3)^2, k[x_o])_o$$

where $t_j(i) = x_j - \alpha_j(i)x_o$ $j = 1,2,3$, $i = 1,\dots,r$ are local coordinates at P_i. Note that $\alpha_3(i) = 0$ for all $i = 1,\dots,r$.

Denote by $\underline{n}_f(P_i)$ and $\underline{n}_{x_3}(P_i)$ (or simply $\underline{n}_f$ and $\underline{n}_{x_3}$) the element of $H^o(Y,N_Y \otimes O_{P_i})$ defined by

$$\underline{n}_f(P_i)(f) = x_o^n \qquad \underline{n}_f(P_i)(x_3) = 0$$

$$\underline{n}_{x_3}(P_i)(f) = 0 \qquad \underline{n}_{x_3}(P_i)(x_3) = x_o$$

Let $l^o = (l_i^o)_{i=1}^r$, $m^o = (m_i^o)_{i=1}^r$ where $l_i^o : H^o(P_i,N_{P_i}) \to H^o(X,N_X \otimes O_{P_i})$ and $m_i^o : H^o(P_i,N_{P_i}) \to H^o(Y,N_Y \otimes O_{P_i})$ are the restrictions of l^o respectively m^o.

Since the plane $x_3 = 0$ cuts X transversally at P_i the ideal $(\mathfrak{a} + (x_3))$ generates the maximal ideal $\underline{m}_{P_i}$ of $O_{\mathbb{P}^3,P_i} = k[x_1,x_2,x_3]_{P}$. From this we deduce that $l_i^o(\underline{v}_i) = 0$ and $m_i^o(\underline{v}_i) = 0$ only when $\underline{v}_i =$

Recall that $l_i^o(\underline{v}_i) = 0$ is equivalent to $\underline{v}_i$ being parallel to the tangent $\underline{t}_X(P_i)$ of X at P_i. Moreover, it is easy to see that for $(l,Q) \in H^o(Y,N_Y)$, with $l \in k[x_o,x_1,x_2]_{(1)}$, $Q \in k[x_o,x_1,x_2]_{(n)}/(f)$ the projection of $n^o(l,Q)$ onto $H^o(Y,N_Y \otimes O_{P_i})$ is given by

$$n_i^o(l,Q) = l(P_i)\underline{n}_{x_3} + Q(P_i)\underline{n}_f .$$

Summing up, we find

$$A^1_{\underline{D}_o}(\underline{D},O_{\underline{D}}) = \{(l,Q) \in H^o(Y,N_Y) \mid l(P_i)\underline{n}_{x_3} + Q(P_i)\underline{n}_f \text{ parallel to } m_i^o(\underline{t}_X(P_i)) \text{ for } i = 1,\dots,r\}$$

$$= \{(l,Q) \in H^o(Y,N_Y) \mid ((P_i)\underline{n}_f - Q(P_i)\underline{n}_{x_3}) \cdot m_i^o(\underline{t}_X(P_i)) = 0,\ i=1,\dots,r\}$$

where the inner product of $H^o(Y,N_Y \otimes O_{P_i})$ is the one making $\{\underline{n}_f,\underline{n}_{x_3}\}$ an orthonormal basis.

As we have seen in § 2 the morphism $\mathfrak{g}$ corresponds at the tangent space level to the morphism

$$\mathfrak{g}^t : A^1_{\underline{D}_o}(\underline{D},O_{\underline{D}}) \longrightarrow A^1_{\underline{D}''_o}(\underline{D}'',O_{\underline{D}''}) = H^o(L,N_L)$$

where $L = V(X_3)$ is the plane containing Y.

Clearly $H^o(L,N_L) = k[x_o,x_1,x_2]_{(1)}$ and it is easily seen that g^t maps (l,Q) onto l.

Suppose g^t is onto, then there is a basis $\{y_j\}_{j=o}^{q}$ of $A^1_{\underline{D}_o}(\underline{D},O_{\underline{D}})$ such that if we put $y_j = (l_j,Q_j)$, $l_j = x_j$ for $j = 0,1,2$. $l_j = 0$ for $j \geq 3$. Since $Q_j(P_i)\underline{n}_{X_3} \cdot m_i^o(\underline{t}_X(P_i)) = 0$ for $j \geq 3$, $i = 1,\ldots,r$, we find $Q_j(P_i) = 0$ for $j \geq 3$, $i = 1,\ldots,r$.

Moreover for any $i = 1,\ldots,r$, the 3 vectors

$$x_j(P_i)\underline{X}_3(P_i) + Q_j(P_i)\underline{n}_f(P_i) \qquad j = 0,1,2$$

are parallel, proving the relations

$$(x_jQ_k - x_kQ_j)(P_i) = 0$$

for $i = 1,\ldots,r$, and all $j,k = 0,1,2$.

Let's pause to prove the following

<u>Lemma</u>. Let $\mathbb{P}^2 = \text{Proj}(R)$, $R = k[x_o,x_1,x_2]$ and consider any closed subscheme $Z \subseteq \mathbb{P}^2$. Let $J(Z) = J \subseteq R$ be the ideal of those forms vanishing on Z. Then the following statements are equivalent:

1) $\text{Tor}_2^R(R/J,k)_{n+2} = 0$

1)' $\text{Tor}_1^R(J,k)_{n+2} = 0$

2) Given any tripple (Q_o,Q_1,Q_2) with $Q_j \in R_n$, $j = 0,1,2$ such that:

$$X_iQ_j - X_jQ_i \in J_{n+1}$$

for all $i,j = 0,1,2$, then there exists a $Q \in R_{n-1}$ such that

$$Q_i - X_iQ \in J$$

for all $i = 0,1,2$.

<u>Proof</u>. Consider the Koszul complex of R, i.e.

$$0 \to R(-3) \to \coprod^3 R(-2) \to \coprod^3 R(-1) \to R \to k \to 0$$

It is an exact sequence of graded R-modules. Tensorizing by R/J

gives us a complex

$$0 \to R/J(-3) \xrightarrow{d_3} \coprod^3 (R/J)(-2) \xrightarrow{d_2} \coprod^3 (R/J)(-1) \xrightarrow{d_1} (R/J) \to 0$$

The second statement of the lemma (2) is equivalent to $\ker d_2^{n+2} = \operatorname{im} d_3^{n+2}$, i.e. to $\operatorname{Tor}_2^R(R/J,k)_{n+2} = 0$. Q.E.D.

Remark. With the notations above one easily proves the following formula

$$\dim_k J(Z)_n = \sum_{i,j=o}^{n} (-1)^i \binom{n-j+2}{2} \cdot \dim_k \operatorname{Tor}_i^R(J(Z),k)_j .$$

$\dim_k \operatorname{Tor}_o^R(J(Z),k)_n$ is the number of "new" generators of $J(Z)$ of degree n, and $\dim_k \operatorname{Tor}_1^R(J(Z),k)_{n+1}$ is the number of "new" relations among generators of $J(Z)$ of degree $\leq n$.

One may also prove the formula

$$\dim_k J(Z)_n = \binom{n+2}{2} - r + \dim_k H^1(\mathbb{P}^2, \tilde{J}(n)) .$$

We are now ready to prove the main theorem of this paper.

Generalized trisecant lemma. With the notations and assumptions abo suppose Y is the only plane curve of degree n containing $Z = X \cap Y$ Suppose further that $\operatorname{Tor}_2^R(R/J(Z),k)_{m+2} = 0$ for $m \leq n$ and that char k is 0 or big enough. Then the dimension of $\operatorname{im} \mathbf{g}$ at $L = \mathbf{g}($ is ≤ 2 unless X is contained in a hypersurface of the form

$$V(f - f_1 x_3 - f_2 x_3^2 - \cdots - f_n x_3^n) .$$

Proof. The assumptions imply the injectivity of $\mathbf{g}^t$, therefore $\dim_k A^1_{\underline{D}_o}(\underline{D}, O_{\underline{D}}) \leq 3$. The lemma is proved if we are able to prove tha $\dim \hat{O}_{I,Y} = 3$ implies that X sits on a surface of the above form. However $\dim \hat{O}_{I,Y} = 3$ implies $O_{I,Y}$ nonsingular, thus $\hat{O}_{I,Y} = k[[y_o^*, y_1^*, y_2^*]]$ where $\{y_o, y_1, y_2\}$ is a basis for $A^1_{\underline{D}_o}(\underline{D}, O_{\underline{D}})$. As above, we may assume $y_j = (x_j, Q_j)$ $j = 0,1,2$. The formalization $I\hat{}$ of $\tilde{I}$ at Y is therefore defined by the ideal

$$(x_3\hat{}, f\hat{})$$

where:

$$x_3^{\wedge} \equiv X_3 + \sum_{i=0}^{2} x_i y_i^* (\mathrm{mod}(y_0^*, y_1^*, y_2^*)$$

$$f^{\wedge} \equiv f + \sum_{i=0}^{2} Q_i y_i^* (\mathrm{mod}(y_0^*, y_1^*, y_2^*)^2)$$

By the lemma we know there exists a $(n-1)$-form f_1 such that $Q_j - x_j f_1 \in (f)$, $j = 0,1,2$.
Therefore

$$f - f_1 x_3 \equiv f^{\wedge} - f_1 x_3^{\wedge} (\mathrm{mod}(y_0^*, y_1^*, y_2^*)^2)$$

Dehomogenize, by putting $x_o = 1$. Then Y singular at P_i implies $f \in \underline{m}_{P_i}^2$, and $m_i^o(\underline{t}_X(P_i)) = \beta_i \cdot \underline{n}_{x_3}$ with $\beta_i \neq 0$. Consequently $Q_j(P_i) = 0$, i.e. $Q_j \in \underline{m}_{P_i}$ for $j = 0,1,2$. Since $x_o \notin \underline{m}_{P_i}$ we find $f_1 \in \underline{m}_{P_i}$, thus

$$f - f_1 x_3 \in \underline{m}_{P_i}^2$$

since $\underline{m}_{P_i} = \mathfrak{a}_{P_i} + (x_3)_{P_i}$ where $\mathfrak{a}_{P_i} = \mathfrak{a} \cdot k[x_1, x_2, x_3]_{P_i}$ and $(x_3)_{P_i} = (x_3) \cdot k[x_1, x_2, x_3]_{P_i}$.
If Y is nonsingular at P_i, the condition

$$(x_o(\underline{n}_f - f_1 \underline{n}_{x_3}) \cdot m_i^o(\underline{t}_X))(P_i) = 0$$

implies

$$f - f_1 x_3 \in \underline{m}_{P_i}^2 .$$

From this we deduce

(1) $$f - f_1 x_3 \equiv \sum_{p,q=o}^{2} Q_{p,q}\, y_p^* y_q^* (\mathrm{mod}((x_3^{\wedge}, f^{\wedge}) + (y_0^*, y_1^*, y_2^*)^3)$$

(2) $$\begin{aligned} f - f_1 x_3 &= a_{2i} + R_{2i}\, x_3^2 \\ &\equiv a_{2i} + R_{2i}(x_o y_0^* + x_1 y_1^* + x_2 y_2^*)^2 (\mathrm{mod}((x_3^{\wedge}, f^{\wedge}) + (y_0^*, y_1^*, y_2^*)^3) \end{aligned}$$

where $a_{2i} \in \mathfrak{a}_{P_i}$, $R_{2i} \in k[x_1, x_2, x_3]_{P_i}$, and where $Q_{p,q} \in k[x_o, x_1, x_2]_{(n)}/(f)$ are supposed to be symmetric in p,q. Here we use the assumption $\mathrm{char}\, k \neq 2$.

As in the proof of the classical trisecant lemma we may assume that the formalization $Z^\wedge$ of $\widetilde{Z}$ at Y as a closed subscheme of $\mathbb{P}^3 \otimes \hat{O}_I$ is defined by the ideal

$$\prod_{i=1}^{r} (x_1-\alpha_1^\wedge(i)x_o, x_2-\alpha_2^\wedge(i)x_o, x_3-\alpha_3^\wedge(i)x_o)$$

where $\alpha_j^\wedge(i) \in \hat{O}_{I,Y} \simeq k[[y_o^*,y_1^*,y_2^*]]$ $j=1,2,3$, $i=1,\dots,r$, and $\alpha_j^\wedge(i) \equiv \alpha_o(i)(\mathrm{mod}(y_o^*,y_1^*,y_2^*))$ $j=1,2,3$, $i=1,\dots,r$.

Evaluate (1) and (2) on $(\alpha_1^\wedge(i),\alpha_2^\wedge(i),\alpha_3^\wedge(i))$ and find

$$\sum_{p,q=o}^{2} Q_{p,q}(P_i)y_p^*y_q^* = R_{2i}(\alpha_1(i),\alpha_2(i),\alpha_3(i))(x_oy_o^*+x_1y_1^*+x_2y_2^*)^2$$

(Remember that $\alpha_3(i) = 0$ and $x_o = 1$.) Thus

$$Q_{p,q}(P_i) = R_{2i}(P_i)\,x_p(P_i)\,x_q(P_i)$$

for all $p,q = 0,1,2$ and all $i=1,\dots,r$. This implies

$$(x_rQ_{p,q} - x_pQ_{r,q})(P_i) = 0$$

for all $p,q = 0,1,2$ and all $i=1,\dots,r$.

By the lemma there exists a Q_q such that $Q_q \in k[x_o,x_1,x_2]_{(n-1)}$ an[d] $Q_{p,q} \equiv x_pQ_q(\mathrm{mod}(f))$. Since $Q_{p,q} = Q_{q,p}$ we find $x_pQ_q \equiv x_qQ_p(\mathrm{mod}(f$[...]
Therefore by the lemma used for $(n-1)$ there exists a $f_2 \in k[x_o,x_1,x_2]_{(n-2)}$ such that $Q_p = x_pf_2$ for all $p=0,1,2$, therefore

$$Q_{p,q} \equiv x_px_qf_2(\mathrm{mod}(f)).$$

In particular we find

$$R_{2i}(P_i) = f_2(P_i) \quad \text{for } i=1,\dots,r.$$

Subtract $f_2 \cdot x_3^2$ from both sides of (1) and (2), then we obtain:

(3) $$f-f_1x_3-f_2x_3^2 \equiv \sum_{p,q,r=o}^{2} Q_{p,q,r}\, y_p^*y_q^*y_r^*(\mathrm{mod}((x_3^\wedge,f^\wedge)+(y_o^*,y_1^*,y_2^*)^4))$$

(4) $$\begin{aligned} f-f_1x_3-f_2x_3^2 &= a_{2i} + (R_{2i}-f_2)x_3^2 \\ &\equiv a_{3i} + R_{3i}\cdot(x_oy_o^*+x_1y_1^*+x_2y_2^*)^3(\mathrm{mod}((x_3^\wedge,f^\wedge)+(y_o^*,y_1^*,y_2^*)^4) \end{aligned}$$

where $a_{3i} \in \mathfrak{a}_{P_i}$, $R_{3i} \in k[x_1,x_2,x_3]_{P_i}$ and where $Q_{p,q,r} \in k[x_o,x_1,x_2]_{(n)}/(f)$ are supposed to be symmetric in p,q,r. Here again we use the assumption char $k > 3$.

Now evaluate (3) and (4) on $(\alpha_1^\wedge(i),\alpha_2^\wedge(i),\alpha_3^\wedge(i))$ and find

$$\sum_{p,q,r=o}^{2} Q_{p,q,r}(P_i)y_p^* y_q^* y_r^* = R_{3i}(P_i)(x_o y_o^* + x_1 y_1^* + x_2 y_2^*)^3 .$$

As above we conclude the existence of an $f_3 \in k[x_o,x_1,x_2]_{(n-3)}$ such that

$$Q_{p,q,r} \equiv x_p x_q x_r \cdot f_3 \,(\mathrm{mod}(f)) .$$

Repeating this process we obtain

$$f - f_1 x_3 - f_2 x_3^2 - \cdots - f_n x_3^n \in (x_3^\wedge, f^\wedge) ,$$

By faithfully flatness of the formalization, we deduce from this

$$f - f_1 x_3 - f_2 x_3^2 - \cdots - f_n x_3^n \in (\tilde{x}_3, \tilde{f}) = \tilde{C} .$$

Therefore, since $X \subseteq \varphi(\tilde{I})$, we find

$$X \subseteq V(f - f_1 x_3 - \cdots - f_n x_3^n) \qquad \text{Q.E.D.}$$

<u>Corollary</u>. Let $X \subseteq \mathbb{P}^3$ be a reduced and irreducible curve of degree d. Assume $n < \sqrt{d}$, then the generic hyperplane section of X contains no finite subscheme Z with

$$J(Z)_n \neq 0 \quad \text{and} \quad \mathrm{Tor}_2^R(R/J(Z),k)_{m+2} = 0 \quad \text{for } m \leq n ,$$

unless X is contained in a surface of degree n.

<u>Proof</u>. Let H be the generic hyperplane and let $Z_o = X \cap H$. We may assume, using Bertini, that $Z_o = \{P_1,\ldots,P_d\}$ is nonsingular and that each point P_i is nonsingular on X.

Suppose $\dim_k J(Z_o)_n \geq 2$. Let $f,g \in J(Z_o)_n$ be linearly independent. Since $d > n^2$, f and g must have a common factor, i.e. $f = f_1 \cdot h_1$, $g = g_1 \cdot h_1$. Assume f_1 and g_1 have no common factors and put

n_1 = degree h_1. Then

$$Z_1 = \{P \in Z_o \mid h_1(P) = 0\}$$

has cardinality $\geq d - (n-n_1)^2 > (1+n_1)\cdot n_1$.

Suppose $J(Z_1)_{n_1}$ has dimension ≥ 2, then we find an h_2 of degre n_2, $h_1 = h_1' \cdot h_2$ such that

$$Z_2 = \{P \in Z_1 \mid h_2(P) = 0\}$$

has cardinality $> (1+n_2)n_2$. Since by the classical trisecant lemma we know that $H \cap X$ contains no trisecant unless X is plane, we wil eventually find a

$$Z_p \subseteq Z_o$$

such that $J(Z_p)_{n_p} = (h_p)$ and $|Z_p| > (n_p+1)n_p$.

Suppose $J(Z_p)_{n_p+1}$ has dimension ≥ 4, pick an element $g \in J(Z_p)_{n_p}$ $g \notin (h_p)$. Since g and h_p have more than $(n_p+1)n_p$ common roots there exists a h_{p+1} of degree n_{p+1} such that $h_p = h_p' \cdot h_{p+1}$, $g = g' \cdot h_{p+1}$. Moreover

$$Z_{p+1} = \{P \in Z_p \mid h_{p+1}(P) = 0\}$$

has cardinality $> (1+n_p)\cdot n_p - (n_p-n_{p+1})(n_p-n_{p+1}+1) > (n_{p+1}+1)n_{p+1}$.

Continuing this process we arrive at a form f of degree $m \leq n$ and subscheme

$$Z_f = \{P \in Z_o \mid f(P) = 0\}$$

such that

$$\dim_k J(Z_f)_m = 1 \qquad \dim_k J(Z_f)_{m+1} = 3.$$

Obviously Z_f and f satisfy the conditions in the Generalized trisecant lemma, therefore there exists a surface of the form

$$V(f - f_1x_3 - \cdots - f_mx_3^m)$$

containing X.

We may therefore assume $\dim_k J(Z_o)_n \leq 1$.

If $\dim_k J(Z)_n \geq 2$ enlarge Z inside Z_o until the enlarged Z_1 ha

$\dim_k J(Z_1)_n = 1$.

In the process we do not introduce new relations among generators of degree $\leq n$, therefore (see the Remark above)

$$\mathrm{Tor}_2^R(R/J(Z_1),k)_{m+2} = 0 \qquad \text{for } m \leq n .$$

Let $J(Z_1)_n = (f)$, then Z_1 and f satisfy the conditions of the G.T.L. and there exists a surface of the form

$$V(f - f_1x_3 - \cdots - f_nx_3^n)$$

containing X. Q.E.D.

<u>Corollary</u>. Suppose $d > (n+1)n$ then the generic hyperplane section of X is not contained in a curve of degree n unless X is contained in a surface of degree n.

<u>Proof</u>. Suppose $Z_o = X \cap H$ is contained in a curve of degree n. Then as in the first part of the proof of the previous lemma, there exists a subscheme Z_f of Z_o such that $J(Z_f)_m = (f)$ and $\dim_k J(Z_f)_{m+1} = 3$, with $m \leq n$. Therefore X is contained in a surface of the form $V(f - f_1x_3 - \cdots - f_mx_3^m)$. Q.E.D.

<u>Corollary</u>. With the assumptions above, suppose the generic hyperplane section $H \cap X$ contains 6 points on a conic, then X sits on a quadric.

<u>Proof</u>. As above, let $Z_o \subseteq H \cap X$ be a subscheme such that $J(Z_o)_2 \neq 0$ and $\mathrm{card}(Z_o) = 6$. By the ordinary trisecant lemma, we know that $J(Z_o)_2 = (f)$. Moreover, if $J(Z_o)_3$ has dimension 5 or more, there must exist two elements $g,h \in J(Z_o)_3$ such that

$$x_of, x_1f, x_2f, g, h$$

are linearly independent. But then one may easily find coefficients $\alpha, \beta \in k$ such that $\alpha g + \beta h$ and f meets in 7 points. This means that f is a union of two lines, which is impossible since no line of

H contains more that 2 points of $H \cap X$.
Therefore $\dim_k J(Z_o)_2 \leq 4$ and one immediately finds $\mathrm{Tor}_2^R(R/J(Z_o),k)_4 = 0$. The conclusion then follows from the theorem. Q.E.D.

<u>Remark</u>. One would be tempted to conjecture, along the lines of the last corollary, that the generic hyperplane section of X contains no $\binom{n+2}{2}$ points on a curve of degree n unless X sits on a surface of degree n. However, the following example, essentially due to Peskine proves that the "conjecture" is false even in degree $n = 3$.

<u>Example</u> (C. Peskine). Let C be a smooth curve of degree 6 and of genus 3 on a smooth quadric (a correspondence (2,4) on the quadric). Let S and S' be two general surfaces of degree 4 containing C. The residual intersection of S and S' is a smooth connected curve X of degree 10 and genus 11, not contained in a cubic surface (use for example the exact sequence

$$0 \to \omega_C(-4) \to O_{S\cap S'} \to O_X \to 0)$$

Every hyperplane section $H \cap X$ is, in H, the residual intersection of two plane quartics containing $H \cap C$. Since $H \cap C$ is the complete intersection of a conic and a cubic plane curve, one shows (see [PS] § 3) that $H \cap X$ admits the following resolution

$$0 \to O_H(-6) \oplus O_H(-5) \to O_H^2(-4) \oplus O_H(-3) \to O_H \to O_{H\cap X} \to 0$$

Therefore $H \cap X$ is always on a plane cubic curve.

Bibliography

[Ab] Abhyankar, S., Algebraic Space Curves, Séminaire de Mathématiques supérieures Été 1970. Les presses de l'Université de Montréal (1971).

[An] Andreotti, Aldo, On a theorem of Torelli. Americal Journal of Mathematics Vol 80 (1958) pp. 801-828.

[E] Ellingsrud, Geir, Sur le schèma de Hilbert des variétés de codimension 2 dans $\mathbb{P}^e$ à cône de Cohen-Macaulay. Annales Sci. de l'Ecole Normale Supérieure. 4e série t. 8 (1975) p. 423-431.

[H] Holme, Audun, Embedding-obstruction for singular algebraic varieties in $\mathbb{P}^N$. Acta mathematica Vol 135 (1975) pp. 155-185.

[K] Kleppe, Jan, Deformation of Graded Algebras. Preprint Series Department of Math., University of Oslo, no. 14 (1975).

[L] Laksov, Dan, Some enumerative properties of secants to non-singular projective schemes. Math. Scand. 39 (1976) pp. 171-190.

[La 1] Laudal, O.A., Sur la théorie des limites projectives et inductives. Théorie homologique des ensembles ordonnés. Annales Sci. de l'Ecole Normale Supérieure. 3e série t. 82 (1965) pp. 241-296.

[La 2] Laudal, O.A., Sections of functors and the problem of lifting (deforming) algebraic structures I. Preprint Series no. 24 (1975) Institute of Mathematics, University of Oslo.

[La 3] Laudal, O.A., Sections of functors and the problem of lifting (deforming) algebraic structures III. Preprint Series no. 6 (1976) Institute of Mathematics, University of Oslo.

[M] Mumford, David, Algebraic Geometry I. Complex Projective Varieties, Springer-Verlag, Berlin, Grundlehren der Mathematischen Wissenschaften 221, (1976).

[P] Peters, C.A.M. and Simonis, J., A secant formula. Quarterly J. Math. (Oxford) 27 (1976) pp. 181-189.

[PS] Peskine, C. and Szpiro, L., Liaison des variétés algébriques. Invent. Math. vol 26 (1974) pp. 271-302.

[Sa] Samuel, P., Lectures on old and new results on algebraic curves. Tata Institute of Fund. Research, Bombay (1966).

Deformations of Curves I

Moduli for Hyperelliptic Curves

by O.A. Laudal and K. Lønsted [1)]

This is an expanded version of the talk delivered by the second author at the Symposium in Algebraic Geometry in Tromsø, 1977.

Fix an algebraically closed field k and an integer $g \geq 2$. Let M_g denote the coarse moduli scheme for smooth curves over k of genus g. There are many natural possibilities of defining subsets of M_g by requiring special geometric properties of the corresponding curves. One is to ask for the existence of Weierstrass points with given gap-sequence. This problem has been studied by Rauch [18], Farkas [4] and more recently by Arbarello [2], who has given precise statements about dimension and existence of the subsets (when $k = \mathbb{C}$).

Another possibility is to require the existence of a morphism of given degree onto an arbitrary curve of given genus. This was initiated by Hurwitz [8]. A complete account of dimension - and existence questions may be found in Lange [9].

Finally one may distinguish curves by the presence of special subgroups of their automorphism groups. This is the problem studied in the present paper. General results in this direction seem to be rather sparse. However, some profound results have been obtained in the case abelian subgroups (of small order), cf. Accola [1]. Our strategy for attacking this problem is the following. One should first work on the fine moduli spaces $M_{g,n}$ (see Section 1 for definitions) and try to define corresponding subsets V of $M_{g,n}$ as representing schemes for appropriate subfunctors $\mathcal{V}$ of the fine moduli functor $\mathcal{M}_{g,n}$. Then one takes quotients $\overline{V}$ with respect to the group $GL(2g,\mathbb{Z}/n)$ (which makes M_g a quotient of $M_{g,n}$). At last one must decide whether or not $\overline{V}$ is naturally embedded in M_g, hence isomorphic to the subset one is looking for (thus endowed with a natural scheme structure).

In this paper we first give a formula for the completed local rings of V (Theorem 1) in terms of group cohomology and a universal

----

1) Partly supported by the Danish National Research Council under grant no. 511-8212.

obstruction homomorphism, obtained from the deformation theory in [10]. An immediate consequence of the formula is the smoothness of V when the order of the subgroup is prime to char(k).

Then we turn to hyperelliptic curves and combine this formula with the global results in [13]. We thus show that the subset V (now called $H_{g,n}$) exists and that it is smooth of dimension $2g-1$ (Theorem 3). We see that the quotient $\overline{V} = H_g$ is a coarse moduli scheme for hyperelliptic curves and give partial results for the natural morphism $H_g \to M_g$ (Prop. 3). The proof of these rely upon the fact that the set of hyperelliptic curves with "many" automorphisms form a closed subset of H_g of codimension $\geqq 2$ (Lemma 3).

The authors should like to express their gratitude to the University of TromsØ, and in particular to prof. Loren Olson, for generous hospitality and for the opportunity of including this report in the Symposium.

1. Automorphism-preserving deformations of curves.

By a curve of genus g over a scheme S we mean a smooth projective morphism $p: C \to S$ whose geometric fibres are irreducible curves of genus g. We let J denote the jacobian functor on the category of curves over S, and for any integer n we let ${}_nJ$ denote the subfunctor of J whose value at a curve C over S is the subgroup-scheme of n-division points on $J(C)$.

Let $g \geqq 2$ and $n \geqq 3$ be integers. For a $\mathbb{Z}[n^{-1}]$-scheme S we set

$$(1.1) \qquad \mathcal{M}_{g,n}(S) = \left\{ (C,\varphi) \,\middle|\, \begin{array}{l} C \text{ curve of genus } g \text{ over } S \\ \varphi \text{ level-n-structure on } C \end{array} \right\} / \sim$$

In this definition a level-n-structure on C is an isomorphism $\varphi: {}_nJ(C) \xrightarrow{\sim} (\mathbb{Z}/n)^{2g}_S$, where $(\mathbb{Z}/n)^{2g}_S$ is the constant group-scheme over S associated with the abelian group $(\mathbb{Z}/n)^{2g}$. Two pairs (C,φ) and (C',φ') are called equivalent if there exists an S-isomorphism $\tau: C \xrightarrow{\sim} C'$ such that $\varphi = \varphi' \circ {}_nJ(\tau)$.

The definition in (1.1) extends to defining a functor $\mathcal{M}_{g,n}$: $\underline{\text{Sch}}_n^{\text{o}} \to \underline{\text{Sets}}$, where $\underline{\text{Sch}}_n$ denotes the category of schemes over $\mathbb{Z}[n^{-1}]$. The group $\Gamma_n = GL(2g, \mathbb{Z}/n)$ acts on $\mathcal{M}_{g,n}$ in an obvious way and the basic results about $\mathcal{M}_{g,n}$ may be formulated as follows.

THEOREM A (Grothendieck-Mumford). $\mathcal{M}_{g,n}$ is representable by a scheme $M_{g,n}$ which is quasi-projective and smooth of relative dimension $3g-3$ over $\mathbb{Z}[n^{-1}]$. $M_{g,n}$ is called the fine moduli scheme for smooth curves of genus g with level-n-structure.

For varying n the quotients $M_{g,n}/\Gamma_n$ glue together to a quasi-projective $\mathbb{Z}$-scheme M_g, which is a coarse moduli scheme for curves of genus g.

One may consult [5 , exposé 195], [15, Chap. 5], [3] or [16] for (hints of) a proof of this theorem. We shall now study $M_{g,n}$ and M_g locally around a point, using the deformation theory in [10] and [11]. This will be done by a study of the fiberfunctors of (the functor defined by) $M_{g,n}$.

Let k be an algebraically closed field and choose the integer $n \geqq 3$ such that $\text{char}(k) \nmid n$. Consider a point $c \in M_{g,n}(k)$ represented by a pair (C,φ). Set $\Phi_c = \{\varphi \circ {}_nJ(\tau) \mid \tau \in \text{Aut}_k(C)\}$. Then another pair (C,φ') represents the same point c if and only if $\varphi' \in \Phi_c$. The fiberfunctor $\mathcal{M}_c$ of $\mathcal{M}_{g,n}$ over c is defined on the category $k/\underline{Sch}/k$ of pointed k-schemes. For a k-scheme S with k-point η: $\mathrm{Spec}(k) \to S$ one has

$$(1.2) \qquad M_c(S) = \left\{ (\overline{C},\overline{\varphi}) \left| \begin{array}{l} \overline{C} \text{ curve of genus } g \text{ over } S \,; \\ \overline{\varphi} \text{ level-n-structure on } \overline{C} \,; \\ \eta^*\overline{C} = C \,;\ \eta^*\overline{\varphi} \in \Phi_c \end{array} \right. \right\} / \sim$$

One has a forgetfull morphism of functors $M_c \to Def_C$, where $Def_C : k/\underline{Sch}/k \to \underline{Sets}$ is the deformation functor of C (see [10]), defined by

$$Def_C(S) = \left\{ \overline{C} \left| \begin{array}{l} \overline{C} \text{ curve of genus } g \text{ over } S \\ \eta^*\overline{C} = C \end{array} \right. \right\} / \sim$$

LEMMA 1. The morphism $M_c \to Def_C$ is smooth and induces an isomorphism on the tangentspaces (i.e., $M_c(k[\varepsilon]) \simeq Def_C(k[\varepsilon])$, where $k[\varepsilon]$ denotes the dual numbers over k).

Proof. For any curve $\overline{C} \to S$ the groupscheme ${}_nJ(\overline{C})$ is étale over S. The first assertion then follows from [6, Exp. I, Thm. 5.5] (or [7, Thm. IV. 18.1.2]). The second assertion is an immediate consequence of the existence of a section to $\mathrm{Spec}\, k[\varepsilon] \to \mathrm{Spec}\, k$. ∎

So, the functors M_c and Def_C have the same hull, H_C. Since M_c is representable by the local ring $o_{c,M_{g,n}}$, H_C is isomorphic to the completion $\hat{o}_{c,M_{g,n}}$. On the other hand it was proved by Grothendieck ([5]) that H_C pro-represents Def_C and that

$$H_C = \mathrm{Sym}_k(H^1(C,\Theta_C)^*)^\wedge ,$$

where Θ_C denotes the tangentsheaf on C. This formula for H_C is a special case of Theorem 1 below. Since $\dim_k H^1(C,\Theta_C) = 3g-3$ this yields the smoothness and the dimension of $M_{g,n}$ over k.

Now, let G denote a group. We consider triples (C,φ,ν), where $C \to S$ is a curve of genus g over a scheme S, φ is a level-n-structure on $\overline{C}$, and $\nu: G \to \mathrm{Aut}_S(C)$ is a grouphomomorphism. Two triples (C,φ,ν) and $(C,,\varphi',\nu')$ are called equivalent if there exist S-isomorphisms $\sigma,\tau: C \simeq C'$ such that $\varphi = \varphi' \circ {}_nJ(\tau)$ and $\nu = \sigma^{-1}\nu'\sigma$ (i.e: $\nu(g) = \sigma^{-1} \circ \nu'(g) \circ \sigma$ for all $g \in G$). We define a functor $M^G_{g,n} : \underline{Sch}^o_n \to \underline{Sets}$ by setting

$$(1.3)\quad M^G_{g,n}(S) = \left\{ (C,\varphi,\nu) \;\middle|\; \begin{array}{l} C \text{ curve of genus } g \text{ over } S\,, \\ \varphi \text{ level-n-structure on } C\,; \\ \nu\colon G \to \mathrm{Aut}_S(C) \text{ grouphomomorphism} \end{array} \right\} / \sim$$

An element of $M^G_{g,n}(k)$ is determined by a pair (c,ν), where is a k-point of $M_{g,n}$ and $\nu\colon G \to \mathrm{Aut}_k(C)$ is a grouphomomorphism. S $N_c = \{\sigma^{-1}\nu\,\sigma \mid \sigma \in \mathrm{Aut}_k(C)\}$ and denote the fiberfunctor of $M^G_{g,n}$ ove the element determined by (c,ν) by $M_{(c,G)}$. Then one has

$$(1.4)\quad M_{(c,G)}(S) = \{(\overline{C},\overline{\varphi},\overline{\nu}) \mid \mathrm{class}(\overline{C},\overline{\varphi}) \in M_c(k);\ \eta^*\overline{\nu} \in N_c\} / \sim$$

for any pointed k-sheme $\eta\colon \mathrm{Spec}(k) \to S$, where $\eta^*\colon \mathrm{Aut}_S(\overline{C}) \to \mathrm{Aut}_k(C)$ is the restriction map induced by η. Let $Def_{(C,G)}$: $k/\underline{Sch}/k \to \underline{Sets}$ be defined by

$$(1.5)\quad Def_{(C,G)}(s) = \left\{ (\overline{C},\overline{\nu}) \;\middle|\; \begin{array}{l} \overline{C} \text{ curve of genus } g \text{ over } S\,; \\ \overline{\nu}\colon G \to \mathrm{Aut}_S(\overline{C}) \text{ grouphomomorphism}\,; \\ \eta^*\overline{C} = C\,;\ \eta^*\overline{\nu} \in N_c \end{array} \right\} /\sim$$

Then we have a natural forgetfull morphism of functors $M_{(c,G)} \to Def_{(C,G)}$, which is injective on the category of <u>connected</u> pointed k-schemes, since the automorphism scheme $\underline{\mathrm{Aut}}_S(\overline{C})$ is unramified over S (see [3]).

LEMMA 2. The morphism $M_{(c,G)} \to Def_{(C,G)}$ is smooth and induces an isomorphism on the tangent spaces.

<u>Proof</u>. Similar to the proof for Lemma 1. ▮

This lemma implies that $M_{(c,G)}$ and $Def_{(C,G)}$ have the same hull, $H_{(C,G)}$. Assume now that the homomorphism $\nu\colon G \to \mathrm{Aut}_k(C)$ is injective, i.e., G may be identified with a subgroup of $\mathrm{Aut}_k(C)$. Let $\underline{G}$ denote the subcategory of $\underline{Sch}/_k$ whose only object is C an whose morphisms are given by the elements of G. It is clear that $Def_{(C,G)}$ is isomorphic to the <u>deformation</u> <u>functor</u> $Def_{\underline{G}}$ (see [11]). Consequently, in this case the hull $H_{(C,G)}$ may be computed from the algebra cohomology of $\underline{G}$. Recall that one has a spectral sequence with 2-term

$$E_2^{p,q} = H^p(G, A^q(k,C,O_C)) = H^p(G, H^q(C,\Theta_C))$$

converging towards $A^i(\underline{G},O_{\underline{G}})$. Since C is a curve of genus $g \geq 2$ this spectral sequence degenerates and one has

$$A^i(\underline{G},O_{\underline{G}}) = H^{i-1}(G, H^1(C,\Theta_C)).$$

The general deformation theory [10] applied to this case yields the following

THEOREM 1. Assume that $\nu: G \to \mathrm{Aut}_k C$ is injective. Then there exists an obstruction morphism $o: T^2 \to T^1$, where $T^i = \mathrm{Sym}_k(H^i(G,H^1(C,\Theta_C))^*)^\wedge$, $i = 1,2$, such that the hull $H_{(C,G)}$ of $M_{(C,G)}$ is given by

$$H_{(C,G)} = T^1 \hat{\otimes}_{T^2} k .$$

In particular, the embedding dimension of $H_{(C,G)}$ is equal to $r = \dim_k H^0(G,H^1(C,\Theta_C))$.

An interesting special case occurs when $\mathrm{char}(k)$ is prime to the order of G.

COROLLARY. Assume furthermore that $\mathrm{char}(k) \nmid |G|$. Then one has an isomorphism $H_{(C,G)} \simeq k[[t_1,\cdots,t_r]]$.

Proof. In this case $H^1(G,H^1(C,\Theta_C)) = 0$. I

So if the assumption of the corollary holds and if $M^G_{n,g}$ (or a suitable subfunctor with the same fiberfunctor as $M^G_{n,g}$ is representable, then the representing scheme is smooth of dimension r over k.

2. Deformations of hyperelliptic curves.

As before, let $g \geq 2$ and $n \geq 3$ be integers and let k denote an algebraically closed field. A (smooth proper connected) curve C over k of genus g is called hyperelliptic if there exists an involution σ on C (called the canonical involution) such that

(2.1.) $C/\{1,\sigma\} \simeq \mathbb{P}^1_k$.

(See section 3 for other equivalent conditions.) The generalization of this concept to a curve $C \to S$ over an arbitrary base scheme is the following.

Definition 1. A curve $C \to S$ of genus g is called hyperelliptic, if there exists an S-involution σ of S such that $C/\{1,\sigma\}$ is a curve of genus 0 over S.

This definition has a sense since C/G is a smooth curve over S for all subgroups $G \subset \mathrm{Aut}_S(C)$, by [13 , Thm. 4.12]. According to the same reference the quotient C/G commutes with base change on S. Consequently, if we set

(2.2) $\quad \mathcal{H}_{g,n}(S) = \{\mathrm{class}(C,\varphi) \in \mathcal{M}_{g,n}(S) \mid C \text{ hyperelliptic}\},$

for a $\mathbb{Z}[n^{-1}]$-scheme S, we get a <u>subfunctor</u> $\mathcal{H}_{g,n}$ of $\mathcal{M}_{g,n}$ with the <u>same fiberfunctor</u> as $\mathcal{M}_{g,n}^{\{1,\sigma\}}$ above a <u>hyperelliptic</u> curve over k (together with a level-n-structure). Therefore, we may apply the results in section 1 to compute the hulls of the fiberfunctors of $\mathcal{H}_{g,n}$.

Let C now denote a hyperelliptic curve over k with canonical involution σ. Set $G = \{1,\sigma\}$ and let $\nu\colon G \to \mathrm{Aut}_k(C)$ be the inclusion. We shall compute $H^1(C,\Theta_C)$ and the action of σ below. One conclusion is that

(2.3) $\quad \dim_k H^0(G,H^1(C,\Theta_C)) = 2g-1 .$

Let φ be a level-n-structure on C and let c be the corresponding element of $\mathcal{M}_{g,n}(k)$. The main result in this section is

THEOREM 2 (Weak form). Let $H_{(c,G)}$ denote the hull of the fiberfunctor of $\mathcal{H}_{g,n}$ above (c,G).

(a) If $\mathrm{char}(k) \neq 2$, then $H_{(c,G)}$ is smooth of dimension $2g-1$ over k.

(b) If $\mathrm{char}(k) = 2$, then the embedding dimension of $H_{(c,G)}$ is $2g-1$.

<u>Proof</u>. An immediate application of Thm. 1 and its Corollary. ∎

The strong form of this theorem simply asserts that the conclusion in (a) also holds when $\mathrm{char}(k) = 2$. Essentially two distinct proofs of the strong form present themselves. One is purely algebraic and goes via the somewhat lengthy explicit computations of the cup-product and the higher Massey products on H^1 that make up the obstruction morphism $o\colon T^2 \to T^1$ in Thm. 1 (see [11]), thus proving that o is trivial, i.e., $H_{(c,G)} \simeq T^1$. The other one, which is given in section 3, is more geometric. It uses the representability of $\mathcal{H}_{g,n}$ to show that $\dim H_{(c,G)} = 2g-1$. Thus o must be trivial, i.e.,

$H_{(c,G)} \simeq T^1$.

Computation of $H^1(C,\Theta_C)$.

Consider the canonical morphism $f: C \to \mathbb{P}^1_k$ of degree 2 and choose coordinates on $\mathbb{P}^1_k$ such that neither 0 nor ∞ are ramification points for f. Then $\mathbb{P}^1_k$ is glued together if two affine lines $\mathrm{Spec}\, k[x_1]$ and $\mathrm{Spec}\, k[x_2]$. According to this, C is the push-out of the following diagram

$$\begin{array}{ccccc} \mathrm{Spec}\, A_1 & & & & \mathrm{Spec}\, A_2 \\ & {}^a\ell_1 \searrow & & \swarrow {}^a\ell_2 & \\ & & \mathrm{Spec}\, A_0 & & \end{array}$$

where $A_i = k[x_i,y_i]/(y_i^2 + a_i(x_i)y_i + b_i(x_i))$, $i = 1,2$, and $A_0 = (A_1)_{x_1}$. Here the a_i are polynomials of degree $\leqq g + 1$ and the b_i are polynomials of degree $2g + 2$. When $\mathrm{char}(k) = 2$ we have $\deg(a_i) = g + 1$. These polynomials may be chosen so that they satisfy the relations

$$a_2(x_2) = x_2^{g+1}\, a_1(x_2^{-1}) ,$$

$$b_2(x_2) = x_2^{2g+2}\, b_1(x_2^{-1}) .$$

(See [13 , section 5] for analogous computations.)

The morphism ${}^a\ell_1$ is induced by the localization map $\ell_1: A_1 \to (A_1)_{x_1}$, and ${}^a\ell_2$ is induced by $\ell_2 : A_2 \to A_0$ defined by $\ell_2(x_2) = x_1^{-1}$, $\ell_2(y_2) = x_1^{-(g+1)} \cdot y_1$. It follows that $H^1(C,\Theta_C)$ may be computed as the derived limit $\varprojlim^{(1)}$ of the diagram

$$(2.4) \qquad \begin{array}{ccccc} \mathrm{Der}_k(A_1,A_1) & & & & \mathrm{Der}_k(A_2,A_2) \\ & d_1 \searrow & & \swarrow d_2 & \\ & & \mathrm{Der}_k(A_0,A_0) & & \end{array}$$

where d_1 and d_2 are induced by ℓ_1 and ℓ_2. Here d_1 is the localization and d_2 may be explicited as follows. For $D \in \mathrm{Der}_k(A_2,A_2)$ one has

$$d_2(D)(x_1) = \ell_2(D(x_2^{-1})) ,$$

$$d_2(D)(y_1) = \ell_2(D(x_2^{-(g+1)} \cdot y_2)) .$$

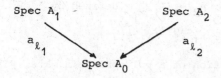

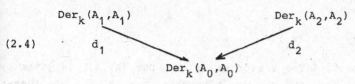

Since C is nonsingular, it is easy to compute $\mathrm{Der}_k(A_i,A_i)$. Identif ing $\mathrm{Der}_k(k[x_i,y_i], k[x_i,y_i])$ with $k[x_i,y_i]^2$ via the basis $\{\frac{\partial}{\partial x_i}, \frac{\partial}{\partial y_i}\}$ one finds

$$
(2.5)\quad \begin{aligned} \mathrm{Der}_k(A_i,A_i) &= \{(\alpha_i,\beta_i) \in A_i^2 \mid \alpha_i(y_i\cdot\frac{da_i}{dx_i} + \frac{db_i}{dx_i}) + \beta_i(2y_i+a_i) = 0 \\ &= A_i\cdot(-2y_i-a_i\ ,\ y_i\cdot\frac{da_i}{dx_i} + \frac{db_i}{dx_i}) \ . \end{aligned}
$$

This identifies $\mathrm{Der}_k(A_i,A_i)$ with A_i as an A_i-module. The map d_1 in (2.4) then becomes the usual localization map $A_1 \to (A_1)_{x_1}$, and the map d_2 becomes a map $d_2\colon A_2 \to (A_1)_{x_1}$ defined by

$$
d_2(r) = x_1^{-(g-1)} r(x_1^{-1},\ x_1^{-(g+1)} y_1) \quad ,
$$

for an element $r \in A_2$ (reduction of a polynomial in $k[x_2,y_2]$). Inserting this interpretation in (2.4) we find that

$$
\begin{aligned} H^1(C,\Theta_C) &= (A_1)_{x_1}/\mathrm{Im}(d_1) + \mathrm{Im}(d_2) \\ &= (A_1)_{x_1}/A_1 + \mathrm{Im}(d_2) \ . \end{aligned}
$$

Let $[r]$ denote the class of an element $r \in (A_1)_{x_1}$ in $H^1(C,\Theta$ One readily sees that the set

$$
(2.6)\quad \{[\frac{1}{x_1}],[\frac{1}{x_1^2}],\dots,[\frac{1}{x_1^{g-2}}],[\frac{y_1}{x_1}],[\frac{y_1}{x_1^2}],\dots,[\frac{y_1}{x_1^{2g-1}}]\}
$$

is k-linearly independent. On the other hand one has $\dim_k(H^1(C,\Theta_C)$ 3g-3. Therefore (2.6) is a k-basis for $H^1(C,\Theta_C)$.

The canonical involution σ induces a dual action on the A_i , also denoted by σ. It is given by

$$
(2.7)\quad \begin{aligned} \sigma(x_i) &= x_i \ , \\ \sigma(y_i) &= -y_i - a_i(x_i) \end{aligned}
$$

(See [13, section 5].) The action of σ on $\mathrm{Der}_k(A_i,A_i$ is given by $\sigma^*D = \sigma^{-1} \circ D \circ \sigma$ for $D \in \mathrm{Der}_k(A_i,A_i)$. When $\mathrm{Der}_k(A_i,A_i)$ is identified with A_i as in (2.5), then the corresponding $\sigma^*\colon A_i \to A_i$ becomes

$$
\sigma^*(r) = -\sigma(r) \ ,
$$

where $\sigma(r)$ is defined by (2.7). The action of σ on the basis (2.6) of $H^1(C,\Theta_C)$ is therefore

$$\sigma^*[\frac{1}{x_1^p}] = -[\frac{1}{x_1^p}],\ p = 1,\cdots,g-2\ ;$$

(2.8)
$$\sigma^*[\frac{y_1}{x_1^q}] = [\frac{y_1}{x_1^q}] + [\frac{a_1(x_1)}{x_1^q}],\ q = 1,\cdots,2g-1.$$

In the remaining discussion we have to distinguish between two cases.

Case 1: $\mathrm{char}(k) \neq 2$.

The vectors

(2.9)
$$\tfrac{1}{2}([\frac{y_1}{x_1^p} + \frac{a_1(x)}{x_1^p}])\ ,\ p = 1,\cdots,2g-1$$

are invariant under σ, and the vectors

(2.10)
$$[\frac{1}{x_1^q}],\ q = 1,\cdots,g-2$$

are antiinvariant under σ. It follows that the vectors in (2.9) form a basis for $H^0(G,H^1(C,\Theta_C))$, which proves the formula (2.3), and that the vectors in (2.10) form a basis for a complement to $H^0(G,H^1(C,\Theta_C))$.

Case 2: $\mathrm{char}(k) = 2$.

In this case, first of all the vectors in (2.10) are invariants under σ. Moreover, a k-linear combination $\Sigma_{p=1}^{2g-1} \lambda_p \frac{y_1}{x_1^p}$ is invariant under σ if and only if

$$\sum_{p=1}^{2g-1} \lambda_p [\frac{a_1(x_1)}{x_1}] = 0.$$

The set of such vectors correspond to the kernel of the linear map $k^{2g-1} \to k^{g-2}$ given by the $(g-2)\times(2g-1)$-matrix

(2.11)
$$\begin{pmatrix} a_{10} & a_{11} & \cdot\ \cdot\ \cdot & a_{1,g+1} & & & 0 \\ & a_{10} & a_{11} & \cdot\ \cdot\ \cdot & a_{1,g+1} & & \\ & & \cdot & & \cdot & & \\ & & & \cdot & & \cdot & \\ 0 & & & & \cdot & & \cdot \\ & & & & a_{10} & a_{11} & \cdot\ \cdot\ \cdot & a_{1,g+1} \end{pmatrix}$$

where the entries are the coefficients in the polynomial $a_1(x_1) = a_{10} + a_{11}x_1 + \dots + a_{1,g+1}x_1^{g+1}$. Since $\mathrm{char}(k) = 2$ we have $a_{1,g+1} \neq 0$, so the matrix in (2.11) has rank $g-2$. Thus the kernel of the linear map has dimension $g+1$.

Consequently, in addition to the $g-2$ vectors in (2.10) we get $g+1$ other invariant vectors. The set of all these is clearly a basis for $H^0(G, H^1(C,\Theta_C))$, which proves the formula (2.3) in this case.

A "general" hyperelliptic curve C has σ as the only non-trivial automorphism (see the following section). For such a curve we m compute the embeddingdimension of the local ring of the correspondin point on the coarse moduli scheme M_g, at least when $\mathrm{char}(k) \neq 2$.

PROPOSITION 1. Assume that $\mathrm{char}(k) \neq 2$. Let $c \in M_g(k)$ correspon to a hyperelliptic curve C for which $\mathrm{Aut}_k(C) = \{1,\sigma\}$. Then the em beddingdimension of the local ring $\mathcal{O}_{c,M_g}$ is equal to

$$3g - 3 + \frac{(g-2)(g-3)}{2}$$

Proof. Set $G = \{1,\sigma\}$. The choice of a level-n-structure φ on C determines an embedding $G \to \Gamma_n$. (Actually, in this case the embedd is independent of φ and the image subgroup is simply $\{\pm 1\}$.) Acco ding to [17] the image of G is equal to the isotropy subgroup of for the corresponding point $\bar{c} = (C,\varphi)$ on $M_{g,n}$. Therefore the nat ral morphism $M_{g,n}/G \to M_g$ is étale at $\bar{c}$, by [6, Exp. V, Prop. 2 This implies that we have an isomorphism of complete local rings

$$\text{(2.12)}\qquad \mathrm{Sym}_k(H^1(C,\Theta_C)^*)^{\wedge G} \simeq \hat{\mathcal{O}}_{c,M_g} .$$

Since $\mathrm{char}(k) \nmid |G|$ we may replace $H^1(C,\Theta_C)^*$ by $H^1(C,\Theta_C)$ (2.12). Denote the vectors in (2.9) by u_p, $p = 1,2,\cdots,2g-1$, and the ones in (2.10) by v_q, $q = 1,\cdots,g-2$. It is obvious that $G^{lin} = \mathrm{Im}(G \to \mathrm{Aut}(^m/m^2))$ and G have the same invariant subspace of $^m/m^2$, where m denotes the maximal ideal of $\hat{\mathcal{O}}_{\bar{c},M_{g,n}}$. Therefore, the common embedding dimension of $\mathcal{O}_{\bar{c},M_{g,n}}$ and its completion is equal to that of

$$\text{(2.13)}\qquad \mathrm{Sym}_k(H^1(C,{}_C))^{\wedge G^{lin}} = k[[u_1,\cdots,u_{2g-1},\{v_iv_j \mid 1 \leq i \leq j \leq g-2\}]].$$

The elements $u_1,\cdots,u_{2g-1},\cdots,v_iv_j,\cdots$ are independent over k, $e = 2g-1 + \binom{g-2}{2} + g-2 = 3g-3 + \frac{1}{2}(g-2)(g-3)$. ∎

This proposition is an alternative to Lemma 11 in [16].

3. The moduli schemes for hyperelliptic curves.

Let $H_{g,n}: \underline{Sch}_n^o \to \underline{Sets}$ be the functor defined by (2.2). The principal results of this section are

THEOREM 2. (Strong form). The hulls of the fiberfunctors of $H_{g,n}$ are smooth of dimension $2g-1$.

THEOREM 3. The functor $H_{g,n}$ is representable by a closed subscheme $H_{g,n}$ of $M_{g,n}$ which is smooth of relative dimension $2g-1$ over any field k of characteristic prime to n.

Proofs. When $g = 2$ we have $H_{2,n} = M_{2,n}$, so the theorems follow from Thm. A in this case. We may therefore assume $g \geqq 3$. The proof of theorem 2 is now done in three steps, and in course of the last one we shall prove the part of Thm. 2 that was omitted in section 2.

Step 1. The representability of $H_{g,n}$.

Here we shall need some equivalent formulations of our definition of a hyperelliptic curve $C \to S$ (Def. 1). For simplicity a curve of genus 0 over S is called a twisted $\mathbb{P}_S^1$. The following conditions are then equivalent for a curve $p: C \to S$ of genus $g \geqq 2$ (see [13, Thm. 5.5]).

(3.1) C is hyperelliptic (i.e., it admits a canonical involution σ such that $C/\{1,\sigma\}$ is a twisted $\mathbb{P}_S^1$).

(3.2) There exists a finite surjective S-morphism of degree 2 of C onto a twisted $\mathbb{P}_S^1$.

(3.3) The image of the canonical morphism $f: C \to \mathbb{P}(p_*\Omega_{C/S}^1)$ is a twisted $\mathbb{P}_S^1$ and it commutes with base change on S.

Let $\pi: C \to M = M_{g,n}$ denote the universal curve of genus g and let D denote the image of the canonical morphism $f: C \to \mathbb{P} = \mathbb{P}(\pi_*\Omega_{C/M}^1)$. The co-morphism of f is denoted by ${}^cf$. Set $Q = \text{Coker}({}^cf)$. We have a commutative diagram of coherent sheaves on $\mathbb{P}$,

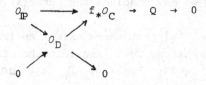

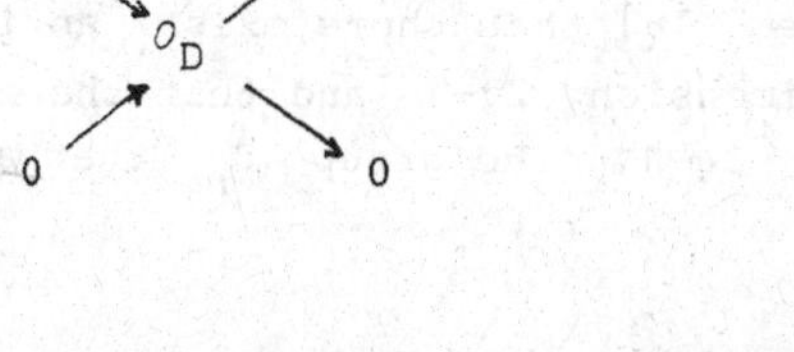

from which we extract the exact sequence

$$0 \to \mathcal{O}_D \to f_*\mathcal{O}_C \to Q \to 0 .$$

For a geometric point x on M with associated curve C_x only two cases may occur in the fibers over x,

(a) either $f_x : C_x \xrightarrow{\sim} D_x$ is an isomorphism and $Q_x = 0$;

(b) or $D_x \simeq \mathbb{P}^1_x$, supp $Q_x = D_x$ and $Q_x \simeq \mathcal{O}_{\mathbb{P}^1}(-g-1)$ on $\mathbb{P}^1_x$. Moreover, the inclusion $D_x \subset \mathbb{P}_x$ is isomorphic to the Veronese morphism $\mathbb{P}^1_x \subset \mathbb{P}^{g-1}_x$.

(See [13, section 5] for this.) Therefore only two distinct Hilbert polynomials for Q may occur in the fibers of $\mathbb{P}$. This implies that the flattening stratification on M for Q (see [14, Lecture 8]) consists of two subschemes, an open M' and a closed subscheme H with support $M-M'$. This stratification has the following property. A morphism $S \to M$ factors through $M' \amalg H \to M$ if and only if the image of the pull-back C_S by the canonical morphism $(= f_S)$ is fla over S and commutes with base change (i.e., if and only if f_S is co-flat, see [13, Prop. 2.13]).

Let $H_{g,n}$ be the one of the two subschemes that correspond to the case (b) above. It follows from (3.3) that $H_{g,n}$ represents the functor $\mathcal{H}_{g,n}$.

Step 2: Dimension and smoothness when $\operatorname{char}(k) \neq 2$.

Since $\mathcal{H}_{g,n}$ is represented by $H_{g,n}$, the fiberfunctors in Thm 2 (Weak form) are pro-represented by their hulls. The assertion about $\dim H_{g,n} \otimes k$ and the smoothness therefore follow from (a) in Thm. (Weak form).

Once $\dim H_{g,n} \otimes k$ is known we see that $H_{g,n} = H$ above. In particular, "most" curves are non-hyperelliptic in genus $g \geqq 3$.

Step 3: Dimension and smoothness when $\operatorname{char}(k) = 2$.

In this case Thm. 2 (Weak form) (b) yields the inequality $\dim H_{g,n} \otimes k \leqq 2g-1$. (So again we have $H_{g,n} = H$.) If we can prove that every irreducible component of $H_{g,n}$ has dimension $2g-1$, ther Thm. 2 (Strong form) hold and we may conclude for $\operatorname{char}(k) = 2$ as in step 2.

It follows from [16, Rem. 16] that there exists an irreducible component of $H_{g,n} \otimes k$ of dimension $2g-1$ and that the other components have dimension at most $2g-1$. The group Γ_n that acts on M_g

maps $H_{g,n}$ onto itself. In order to conclude we only need to show, that Γ_n acts transitively on the set of irreducible components of $H_{g,n} \otimes k$ or, equivalently, that the quotient $(H_{g,n} \otimes k)/\Gamma_n$ is irreducible. This will be proved below for all characteristics (Lemma 1) thus ending the proofs of Thm. 2 and Thm. 3. ∎

For varying n the quotients $H_{g,n}/\Gamma_n$ paste together into a $\mathbb{Z}$-scheme H_g. Let $\mathcal{H}_g$: <u>Sch</u>^{o} → <u>Sets</u> be the functor that to a scheme S associates the set

(3.4) $$\mathcal{H}_g(S) = \{C \to S \mid C \text{ hyperelliptic curve of genus } g\} / \sim .$$

There is a natural morphism $\Phi: \mathcal{H}_g \to H_g$ (i.e., from $\mathcal{H}_g$ to $\mathrm{Hom}(-,H_g)$). For a locally noetherian scheme S it is first defined on the open subschemes $S[p^{-1}]$, p natural prime, using the existence of $H_{g,n}$ for $p \nmid n$. We leave the details to the reader. A <u>coarse moduli scheme</u> for $\mathcal{H}_g$ is defined as in [15, Def. 5.6].

PROPOSITION 2. The $\mathbb{Z}$-scheme H_g (together with Φ) is a coarse moduli scheme for $\mathcal{H}_g$ (i.e., H_g is a coarse moduli scheme for hyperelliptic curves of genus g.

<u>Proof</u>. Similar to the analogous statement for $A_{g,d,n}$ in [15, Thm. 7.10]. ∎

For every $n \geqq 3$ we have a commutative diagram

(3.5) $$\begin{array}{ccc} H_{g,n} & \xrightarrow{\ i\ } & M_{g,n} \\ \downarrow & & \downarrow \\ H_g & \xrightarrow{\ j\ } & M_g \end{array}$$

The vertical arrows are the natural projections, i is the (closed) embedding, and j is the induced morphism. It follows by standard invariant theory that j is finite.

<u>Definition 2</u>. The topological image of j endowed with the reduced scheme structure is denoted by H'_g and called the <u>hyperelliptic locus</u> (in M_g).

LEMMA 1. Let k be a field. Then $H'_g \otimes k$ and $H_g \otimes k$ are irredu-

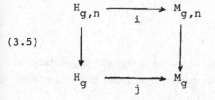

cible of dimension $2g-1$.

<u>Proof</u>. It was proved in [12] that $H'_g \otimes k$ (called H_g in the cite paper) is irreducible. By invariant theory the morphism j in (3.5) induces a homeomorphism of H_g onto H'_g. Hence $H_g \otimes k$ is also irreducible. This was the missing point in the proof of Thm. 3. Now, <u>by</u> Thm. 3 the dimension of $H_g \otimes k$ is $2g-1$. So is that of $H'_g \otimes k$.

For $g \geqq 3$ a natural question in connexion with (3.5) is the following: Is j a closed embedding? We only have a partial answer to this. Let k be an algebraically closed field of characteristic For $p = 2$ then $j \otimes 1_k$ is <u>not</u> an embedding, due to the wild ramification of the map $H_{g,n} \otimes k \to (M_{g,n} \otimes k)/\{\pm 1\}$. Assume then that $p \neq$ If there exists an $n \geqq 3$ such that $p \nmid |\Gamma_n|$, then $j \otimes 1_k$ is a closed embedding. This is a standard result from invariant theory. In terms of cohomology it stems from the vanishing of $H^1(\Gamma_n, M)$, where M is any k-vectorspace.

It seems plausible that $j \otimes 1_k$ is a closed embedding for all fields k with $\mathrm{char}(k) \neq 2$, i.e. $H_g \otimes k \simeq H'_g \otimes k$. The following result gives some support to this statement.

PROPOSITION 3. The morphism $j\colon H_g \to M_g$ is a closed embedding at all points $x \in H_g$ that satisfy

(i) $\mathrm{char}\ \kappa(x) \neq 2$,

(ii) $\dim \mathcal{O}_{x,H_g} \leqq 1$.

The proof is based upon two lemmas.

LEMMA 2. Let C denote a hyperelliptic curve over an algebraically closed field and let $G \subset \mathrm{Aut}(C)$ be a subgroup. Then the quotient C/G is hyperelliptic, elliptic or rational.

<u>Proof</u>. Let σ denote the canonical involution on C. Then σ belongs to the center of $\mathrm{Aut}(C)$. Assume first that $\sigma \in G$ and set $\overline{G} = G/\{1,\sigma\}$. By Galois theory we have a commutative diagram

$$\begin{array}{ccc} C & \longrightarrow & C/\{1,\sigma\} \simeq \mathbb{P}^1 \\ & \searrow \quad \swarrow & \\ & C/G \simeq \mathbb{P}^1/\overline{G} & \end{array}$$

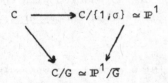

where $\mathbb{P}^1/\overline{G} \simeq \mathbb{P}^1$.

Assume then that $\sigma \notin G$ and let G' be the subgroup of $\mathrm{Aut}(C)$ generated by G and σ, i.e., $G' = G \cup \sigma \cdot G$. Set $\overline{G} = G'/\{1,\sigma\}$. Then G is normal in G' and we have a commutative diagram

$$\begin{array}{ccc} C & \longrightarrow & C/\{1,\sigma\} \simeq \mathbb{P}^1 \\ \downarrow & & \downarrow \\ C/G & \longrightarrow & \mathbb{P}^1/\overline{G} \simeq \mathbb{P}^1 \end{array}$$

which makes C/G a double covering of $\mathbb{P}^1$. This proves the lemma. ▮

LEMMA 3. Let $x \in H_g$ be a point and let C_x denote the corresponding curve over $\overline{\kappa(x)}$. Assume that $\mathrm{Aut}(C_x) \neq \{1,\sigma\}$. Then $\dim \mathcal{O}_{x,H_g} \geqq 2$.

<u>Proof</u>. One may proceed as in the proof of the analogous statement for $x \in M_g$ (see [16, Prop. 14]), except that one replaces $M_{h,n}$ by $H_{h,n}$ according to Lemma 2. ▮

<u>Proof of Prop. 3</u>. Let $x \in H_g$ satisfy (i) and (ii). By Nakayama's lemma we may replace H_g and M_g by $H_g \otimes k$ and $M_g \otimes k$, where k is the algebraic closure of the prime field contained in $\kappa(x)$. Let $n \geqq 3$ be prime to $\mathrm{char}(k)$ and form the diagram (3.5) tensored with k. The subgroup $\{\pm 1\}$ acts trivially on $H_{g,n}$. This yields a new commutative diagram

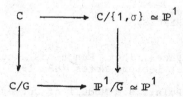

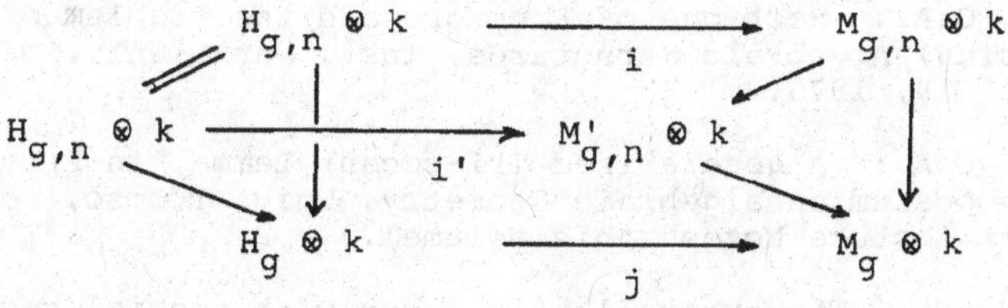

where $M'_{g,n} = M_{g,n}/\{\pm 1\}$ and $M_g = M'_g/\Gamma'_n$, $\Gamma'_n = \Gamma_n/\{\pm 1\}$. Since $\mathrm{char}(k) \neq 2$ the morphism i' is a closed embedding. Since $\mathrm{Aut}(C_x) = \{1,\sigma\}$, the isotropy subgroup of Γ'_n for a point y in $H_{g,n} \otimes k$ above x vanish, i.e., the morphisms $H_{g,n} \otimes k \to H_g \otimes k$ and $M'_{g,n} \otimes k \to M_g \otimes k$ are étale at y, resp. $i'(y)$. Since i' is a closed immersion at y the same follows for j at x. ▮

One may actually prove that there exists an <u>open</u> subset U of $H_g \otimes k$ $(\mathrm{char}(k) \neq 2)$ such that

(i) For all $x \in U$, $\mathrm{Aut}(C_x) = \{1,\sigma\}$;

(ii) $j \otimes 1_k : U \to H'_g \otimes k$ is an open embedding;

(iii) $\mathrm{codim}(H_g \otimes k - U, H_g \otimes k) \geqq 2$.

REFERENCES.

1. Accola, R.D.M.: Riemann Surfaces, Theta Functions, and Abelian Automorphisms Groups, Springer Lecture Notes 483, 1975.

2. Arbarello, E.: Weierstrass Points and Moduli of Curves, Composi tio Math. 29 (1974), 325-342.

3. Deligne, P. and Mumford, D.: The irreducibility of the space of curves of given genus, Publ. Math. de l'I.H.E.S. 36 (1969), 75-

4. Farkas, H.M.: Special Divisors and Analytic Subloci of Teichmül Space, Amer. J. Math. 88 (1966), 881-901.

5. Grothendieck, A.: Fondéments de la Géométrie Algébrique (FGA), traits du Sém. Bourbaki 1957-62, Secrétariat Mathématique, Paris 1962.

6. Grothendieck, A.: Séminaire de Géométrie Algébrique 1960/61 (SGA 1). Revêtements Etale et Groupe Fondamental, Springer Lectu Notes 224, 1971.

7. Grothendieck, A. et Dieudonné, J.: Eléments de Géométrie Algébrique, Chap. IV^4, Publ. Math. de l'I.H.E.S. 32 (1967).

8. Hurwitz, A.: Über Riemann'sche Flächen mit gegebenen Verzweigungspunkten, Math. Ann. 39 (1891), 1-61.

9. Lange, H.: Kurven mit rationaler Abbildung, Göttingen Habilitationsschrift, 1975.

10. Laudal, O.A.: Sections of Functors and the Problem of Lifting (Deforming) Algebraic Structures, Inst. Math. Univ. Oslo Preprin Ser. nr. 18, 1975.

11. Laudal, O.A.: A generalized Tri-secant Lemma, in Proceedings of a Symposium on algebraic Geometry, Univ. Tromsö, July 1977, Springer Lecture Notes (this volume).

12. Lønsted, K.: The hyperelliptic locus with special reference to characteristic two, Math. Ann. 222 (1976), 55-61.

13. Lønsted, K. and Kleiman, S.L.: Basics on Families of hyperellip tic curves, Københavns Univ. Mat. Inst. Preprint Ser. 1977 no. 7 March 1977.

14. Mumford, D.: Lectures on Curves on an Algebraic Surface, Annals of Math. Studies 59, Princeton Univ. Press, 1970.

15. Mumford, D.: Geometric Invariant Theory, Ergebnisse d. Math., Bd. 34, Springer Verlag 1965.

16. Oort, F.: Fine and Coarse Moduli Schemes are different, Report 74-10, Dept. Math., Univ. of Amsterdam, 1974 (partly published in Séminaire d'Algèbre Paul Dubreil Paris 1975-76, Springer Lecture Notes 586, 1977).

17. Popp, H.: The singularities of the moduli scheme of curves, J. Number Theory 1(1969), 90-107.

18. Rauch, H.E.: Weierstrass Points, Branch Points, and the Moduli of Riemann Surfaces, Comm. Pure a. Appl. Math. 12 (1959), 543-560.

O.A. Laudal
Universitetet i Oslo
Matematisk Institut
Blindern - Oslo 3
Norway

K. Lønsted
Matematisk Institut
Universitetsparken 5
2100 København Ø
Denmark

Added in proof: We are very grateful to prof. S. Koizumi for having pointed out that the study of points P in M_g with a given subgroup G of $Aut(C_P)$ has already been undertaken in the analytic case. A fairly algebraic approach may be found in A. Kuribayashi: On analytic families of compact Riemann surfaces with non-trivial automorphisms, Nagoya Math. J. 28 (1966), 119-165, for cyclic groups G of prime order, where the analiticity and the dimension of the corresponding subset of M_g is established.

Later, we have discovered that S. Kravetz has given even more complete results, for general subgroups G , in connexion with his *) solution of the Hurwitz-Nielsen Realization problem, appeared in "On the geometry of Teichmüller spaces and the structure of their modular groups", Ann. Acad. Sci. Fenn. A.I., 278 (1960).

*) incorrect

Double Coverings and Surfaces of General Type

Ulf Persson

Introduction

For any algebraic variety X, we can define the pluri-canoni-cal ring as the graded ring ,

$$R = \Sigma H^{o}(X, mK)$$

R is a birational invariant and $tr.deg_k R \leq \dim X + 1$ .If equality holds,we say that X is of general type (or of hyperbolic type,cf[13]) In this case X is birational to Proj(R),but unless R is finitely gene rated graded Noetherian ring,this does not give a projective model fo X.

We can also define the graded ring,

$$R^{[n]} = \Sigma H^{o}(X, nK)^{\otimes m}$$

which clearly is a finitely generated graded ring,and let $X^{[n]}$=Proj(R It is clear that $X^{[n]}$ is the image under the rational map Φ_{nK} corres ponding the the n-canonical linear system $|nK|$,and is consequently called the n-canonical image.

For a given X , $X^{[n]}$ is birational to X if $n >> 0$.

This work has been partially supported by the National Science Foundation under Grant No. MCS77-07660.

By the general - asymptotic principle we expect $X^{[n]}$ to be birational to X ,for"almost all" varieties for all n. The exceptions are quite interesting to pin down,as they are more special,and thus more amendable to an intimate study.

X is a curve of general type iff its genus is at least two. We have the following list of exceptions to $X^{[n]}$ being birational to X:

$n \geq 3$ No exceptions

$n = 2$ $g = 2$. The bicanonical map is a double cover of $\mathbb{P}^1$.

$n = 1$ X is hyperelliptic (i.e a double cover of $\mathbb{P}^1$)

The canonical map factors through the double cover.

For surfaces the list is a bit more complicated.It is mainly due to Kodaira[7],Moisezon[8] and Bombieri[2].(The latest touches have been given by Bombieri,Catanese,Kulikov and Miayoka)

$n \geq 5$ No exceptions

$n = 4$ $K^2=1$, $p_g=2$

$n = 3$ $K^2=1$, $p_g=2$; $K^2=2$, $p_g=3$

$n = 2$ Surfaces fibered with genus two curves.(If $p_g \geq 6$, $K^2 \geq 10$ these are the only exceptions) and some quite sporadic examples.

$n = 1$ Surfaces which are double covers of surfaces with $p_g=0$. And sporadic examples.

As we will see later,the exceptions for n=3,4 are completly characterized.When n=2 and in particular when n=1,the situation is more chaotic.

For dimensions 3 and higher, not very much is known (see [13]
1. Is the canonical ring finitely generated ? (For surfaces this was shown by Mumford [9])
2. Does there exist a universal n (depending only on the dimension), such that $X^{[n]}$ is birational to n ? If so, find the best possible n!

We also note that for curves the multicanonical maps never have fixed components; the same is true for surfaces provided $n \geq 4$. But this is not the case for higher dimensions (see [13]).

In this article we shall deal exclusively with surfaces, although many of the techniques and propositions carry over ad verbat to the higher dimensional cases. In particular I will indicate a more systematic study of double coverings of surfaces, and how they fit int the exception list above.

As there does not seem to be any standard reference to the elementary facts about double coverings, I will insert one chapter entirely devoted to these elementary considerations (cf also [4], [5] and [12]).

This is a very preliminary article, and the author hopes to extend these investigations in the future.

The neccessary limitations on length will account for occassional terseness and abruptness in the presentation.

I Surfaces of General Type

There is a multitude of invariants for surfaces.

- the Hodge numbers $h^{pq} = \dim H^q(X, \Omega^p)$,where $p_g=h^{20}$,$q = h^{01}$
- the Chern numbers c_1^2 $(=K^2)$, c_2 (the Euler characteristic)
- the holomorphic Euler characteristic. χ

These are of course not independant,and the various relationships are too well-known to be spelled out.suffices it to recall the most basic and deepest - the Noether Formula

$$c_1^2 + c_2 = 12\chi \qquad (\chi = p_g-q+1)$$

Naturally, a most important question is, what invariants can occur,and to what extent the surface is determined by its invariants.

As our basic invariants we choose χ, c_1^2 .Note that to a given pair (χ, c_1^2) there is at most a finite number of families of surfaces with those invariants.(Consider all five-canonical embeddings of surfaces into projective space,then their Hilbertpolynomials are uniquely determined by the two invariants ,and we are done,consult [10])

The invariants (χ, c_1^2) satisfy three basic inequalities.

I. $\chi > 0$, $c_1^2 > 0$

This is classical,and due to,I think Castelnouvo. The example of the Godeaux surface [3] with $\chi=c_1^2 = 1$,shows that these are sharp.

II. $c_1^2 \geq 2\chi - 6$

In fact we have the apparently stronger $c_1^2 \geq 2p_g - 4$.(This was claimed by Castelnouvo,for a modern and elementary prrof see [4]) Horikawa classified in [5] those surfaces for which equality hold.

III $c_1^2 \leq 9\chi$

This is a very deep result,long conjectured and finally proved by Yau (and Miayoka).That this is a best possible inequality has been known for twenty years. Consult [15] where Hirzebruch and Borel give examples of compact qoutients of the open ball in $\mathbb{C}^4$.Those surfaces attain equality in III, by the Hirzebruch proportionality theorem.(It is an open question if such surfaces exist when $\chi=1$ i.e with $p_g=q=0$,$K^2=9$)

We can thus get a rough idea of the totality of all surfaces of general type by plotting the admissable (χ,c_1^2) graphically.

I do believe it is unknown whether these three inequalities completly characterizes the admissable invariants.

Note: If $\pi_1(X)$ contains a finite factorgroup of order m,then II can be sharpened to $c_1^2 \geq 2\chi - 6/m$ .In particular if $q>0$, m can be chosen arbitritarily large,so we have $c_1^2 \geq 2\chi$ for irregular surfaces. Little seems to be known about infinite fundamental groups of surfaces. It has been conjectured that they always contain non-trivial finite factor groups.If that would be the case we would have $c_1^2 \geq 2\chi -3$ for non simply connected surfaces.(Horikawa shows by other means that $c_1^2 \geq 2\chi$- in this case ,see [5])

I would like to close this chapter by mentioning two other inequalities.

Proposition 1.1. If the canonical map is m:1 onto its image then $c_1^2 \geq m(p_g-2)$ (which implies $c_1^2 \geq m(\chi-3)$)

Proposition 1.2. If the canonical map is birational then $c_1^2 \geq 3p_g -7$

For proofs consult [5] and [6] respectively.

Thus the sector $2p_g-4 \leq c_1^2 \leq 3p_g-8$ consists entirely of surfaces for which $X^{[1]}$ is not birational to X.

II Double Coverings

Although the discussion below is valid for varieties of any dimension,we will in order to fix ideas,restrict to the case of surfaces.

Let X,Y be two (non-singular) surfaces,and let $\pi:X\to Y$ be a finite map,generically 2:1.X is then said to be a double cover of Y.

If $\pi:X\to Y$ exhibits X as a double cover of Y,then π induces a unique involution on X i.e a unique action of Z_2 on X.Assume that Y is non-singular, which we will assume from now on,then the reduced image of the fixed point locus of the action,is a curve on Y- the so called branch curve of the double cover.We have $Y = X/Z_2$.

The involution naturally extends to an action on $\mathcal{O}_X$ the structure sheaf of X ,and $\mathcal{O}_X$ splits up into two eigenspaces,correspon ding to the eigen values ±1 . Thus

$$\mathcal{O}_X = \mathcal{O}_X^+ \oplus \mathcal{O}_X^- \quad , \text{ where } \mathcal{O}_X^+ = \pi^*\mathcal{O}_Y \text{ (as subsets of } \mathcal{O}_X\text{)}$$

The multiplicative structure on $\mathcal{O}_X$,turns $\mathcal{O}_X^-$ into an invertible $\mathcal{O}_X^+$ (and hence $\mathcal{O}_Y$) module.It also defines an embedding of $\mathcal{O}_X^- \otimes \mathcal{O}_X^-$ into $\mathcal{O}_X^+$ .Thus if we write $\mathcal{O}_X^- = \mathcal{O}_Y(-B)$ we are given a unique effective divisor of 2B ,this divisor is the branch curve.

Conversely, if we choose a line bundle B and an effective divisor C in $|2B|$, we can define a ring structure on $\mathcal{O}_Y \oplus \mathcal{O}_Y(-B)$, and recapturing X via Spec$(\mathcal{O}_Y \oplus \mathcal{O}_Y(-B))$.

Note:B is just a linebundle defined up to linear equivalence but C is a specific curve.Also note that C does not determine B unique ly.(consult [14] for classical remarks on that fact)

A more geometric (and instructive) construction is given as follows:

Let $[B]\to Y$ be a line bundle with fiber coordinates z_i, $z_i = \phi_{ij} z_j$, let F_i ; $F_i = \phi_{ij}^2 F_j$ be a section of the line bundle $[2B]\to Y$, with zero-locus our branch curve to be C. Consider the subvariety X of [B] given by $z_i^2 - F_i = 0$. The projection $[B]\to Y$ induces X as a double cover of Y, with branch locus C. The involution on X is given by $z_i \to -z_i$. The functions on X can be written as $A_i + z_i B_i$, with ringstructure given by $z_i^2 = F_i$.

This representation yields ;

X is non-singular iff the branch locus is non-singular

X is normal iff the branch locus has no multiple components.

If $\pi: X\to Y$ is a double cover between non-singular surfaces, then π induces maps $\pi^*\mathrm{Div}Y\to \mathrm{Div}X$, and $\pi^*: \mathrm{Pic}Y\to\mathrm{Pic}X$. (The second map is injective, unless $B\neq 0$, $2B=0$, in which the kernel consists of $(0,B)$)

The following proposition and its corollary should be clear.

<u>Proposition 2.1.</u> Let $\pi: X\to Y$ be a double cover, branched at C $(=2B)$, let D be a line bundle on Y , then

i) $\pi_*\pi^*D = \mathcal{O}(D) \oplus \mathcal{O}(D-B)$

ii) $h^i(\pi^*D) = h^i(D) + h^i(D-B)$

iii) $K_X = \pi^*(K_Y + B)$

iv) $(\pi^*D, \pi^*D') = 2(D,D')$

<u>Corollary 2.2.</u> We have the following relationships between the invariants.

i) $K_X^2 = 2(K_Y+B)^2 = 2K_Y^2 + 8(g(B)-1) - 2B^2$

ii) $c_2(X) = 2c_2(Y) + 4(g(B)-1) + 2B^2 = 2c_2(Y) + 2g(C) - 2$

iii) $\chi(X) = 2\chi(Y) + g(B)-1$

iv) $p_g(X) = p_g(Y) + h^0(K_Y+B)$ v) $q(X) = q(Y) + h^1(-B)$

Consider a surface Y with structure sheaf $\mathcal{O}_Y$. We can associate, the function field K(Y) to Y, by letting for each affine open U, $K(Y)(U) = Q\Gamma(U,\mathcal{O}_Y)$ (where Q denotes field of fractions) Note K(Y) is a constant sheaf. (Any surface Y' such that K(Y') = K(Y), is said to be a model for the functionfield K(Y))

Now consider a degree two extension of K(Y) : $K = K(Y)[\Xi]$ where $\Xi^2 = F$, F an element of K(Y). We can now define the normalization X of Y in K. (For each affine U = SpecR ($R = \Gamma(U,\mathcal{O}_Y)$) define V = SpecS where S is the integral closure of R in K. The V' s glue together to form a global surface X.

We have thus constructed a morphism $\pi: X \to Y$, which is a double covering branched along a curve C. Note that X is normal (by construction) but not neccessarily non-singular; hence C may be singular, but it has no multiple components. (C is "square-free")

The following should be clear. (consult [4])

<u>Proposition 2.3.</u> Let Y,Y' be two non-singular models for the function field K(Y), and let X and X' denote their respective normalizations, in K Denote by C and C' the two branch curves. Assume that $p: Y' \to Y$ is a blow up at a point p, with exceptional divisor E_p. Then ,

i) p induces $p': X' \to X$

ii) $C' = p^*C - 2[m_p/2]\, E_p$; where m_p is the multiplicity of C at p and [*] denotes greatest integer.

Note: The key concept in the proof is "square-free"

As a consequence we have the following, again consult [4]

<u>Proposition 2.4.</u> If $\pi: X \to Y$ is a double cover of Y, along a possible singular branch locus, we can find a sequence of blow ups $p: Y' \to Y$, desingularizing C to C'. In other words we get the commutative diagram.

$$\begin{array}{ccc} \pi' : X' & \to & Y' \\ \downarrow & & \downarrow \\ \pi : X & \to & Y \end{array}$$

with X', Y' nonsingular π' a double covering

Note: This gives a very explicit way of desingularizing singularities of double coverings.But space does not allow a digression at this poi Suffices it to state the following.

Proposition 2.5. Let X be a double cover of Y.Then the singular poin of X are rational double points iff the branch locus has no worse si gularities than double points and ordinary triple points (i.e points at which after a blow up,no triple points remain on the curve ,i.e th are no infinitely close triple points).

For a proof see [4]. Note also that the above condition characterizes those points p (singular) ,on C (=2B) ,for which $K_{Y'}+B' = p^*(K_Y+B)$ where p:Y' Y is the blow up at p.(this is essentially the proof)

Recall Brieskorns result on simultaneous resolutions of rational double points, which shows that these singularities can be cor sidered as inessential.(Branchcurves with those singularities can be treated as if they were smooth ! This will be made use of later)

This might be the place to make the following observation.

Proposition 2.6. Exceptional divisors of the first kind on X, occur i two different ways.

i)Pullbacks of exceptional divisors on Y disjoint from the branch loc -these always come in pair.

ii)They are the reduced part of pullbacks from rational components of the branchlocus with selfintersection -2.

Note:Exceptional divisor of the second kind are a bit harder,but only if X is ruled. - The exceptional divisors of type ii) occur as the r solution of isolated fix points of an involution.

The following elementary lemma,will be used later.

Lemma 2.7.Let Y be a non-ruled surface,and let C be a smooth curve wh is even, i.e C (= 2B).Let F be irreducible and assume (K+B)F <0 ,ther

F is a non-singular rational curve, and two cases occur.

i) $F^2 = -1$, F disjoint from C ii) $F^2 = -2$, F component of C.

Proof: Let X be the double cover of Y branched at C. Then X is also non-ruled. Let $\tilde{F} = \pi^*F$, we have $K_X\tilde{F} = 2(K+B)F < 0$, thus F exceptional, and we can apply Proposition 2.6.

It might also be instructive with a direct proof: If $KF < 0$ we have i). If $KF \geq 0$, then $BF < 0$, hence $CF < 0$, thus F component of C, as C smooth $(C-F)F = 0$, i.e $F^2 = CF$. Now $-2 \leq KF+F^2 \leq 2KF+F^2 = 2(K+B)F \leq -2$ and we are done.

<u>Corollary 2.8.</u> Let X be a double cover of Y, X non-ruled. Let L be the free part of K+B. then $2L^2 \leq c_1^2(X)$.

Proof: Write $(K+B)^2 = L^2 + (K+B)F + LF$, and apply above lemma. See also [5] for a more general case.

We conclude this chapter with the following proposition.

<u>Proposition 2.9.</u> Let $\pi : X \to Y$ be a morphism between two projective surfaces, with X non-singular.

Assume that π is generically 2:1, but not neccessarily finite We can then find a commutative diagram.

$$\begin{array}{ccc} X' & \xrightarrow{\pi'} & Y' \\ E \downarrow & & \downarrow \Psi \\ X & \xrightarrow{\pi} & Y \end{array}$$

With X', Y' non-singular, π' finite generically 2:1, and E a sequence of blow ups,

Proof: Desingularize Y to Y", take its normalization X" in K(X), and apply proposition 2.4 !

Note: This resolution only works for 2:1 morphisms, for higher degrees we cannot in general get both X' and Y' non-singular.

The morphism E comes from resolving the isolated fix points of π (cf prop.2.6.ii))

III Multicanonical maps

The following observation plays a pivotal role.

<u>Proposition 3.1.</u> Let X be a double cover of Y, with C =2B the branch locus. Let M be a line bundle on X, which is a pullback from a line bundle H on Y . ($M = \pi^*H$). Then Φ_M factors via Y iff the following holds.

a) $h^0(H-B) = 0$ This happens exactly when the fixed component of M is a pull-back of an effective divisor on Y

b) $h^0(H) = 0$ This happens exactly when the fixed component of M is <u>not</u> a pull-back of an effective divisor on Y

In case a) we have $\Phi_M = \Phi_H \pi$, in case b) $\Phi_M = \Phi_{H-B}\pi$

If X is an unramified cover of Y, then H is not uniquely defined. we th stipulate that H is chosen such that e.g h^0 is maximal. Note also tha in the unramified case, only a) can occur.

Proof: Let L be the free part of M, clearly $L = \pi^*H_0$;where $H_0 = p^*(\mathcal{O}_{\mathbb{P}}$ and $p\pi$ is the factorization of $\Phi_M = \Phi_L$. Thus the fixed part F of M is a pullback of a divisor D on Y, $F = \pi^*D$. By prop.2.1. $1=h^0(F)=h^0(D)+h^0(D-B)$, thus either D or D-B is effective (not both). Furthermore $h^0(H_0) + h^0(H_0-B) = h^0(L) = h^0(M) \leq h^0(H_0)$ again by prop.2.1. Thus $h^0(H_0) = h^0(L) = h^0(M)$ i.e $\Phi_M = \Phi_L = \Phi_{H_0}$,and $h^0(H_0-B) = 0$, finally if D is effective i.e case a) then $h^0(H_0) \leq h^0(H_0+D) \leq h^0(M)$ & $h^0(H_0+D-B)$ if D-B is effective -case b) then $h^0(H_0) \leq h^0(H_0+D-B) \leq h^0(M)$ & $h^0(H_0+D)=$ Observe that H_0 is the free part of H respectively H-B , thus the fin statements. Naturally the argument is reversible.

Note: Case a) is the by far most common case, and case b) is very spec but it does occur; e.g let A be an abelian surface $\pi(z) = -z$, and K the qoutient-the Kummer surface.Let A' be the blow up of A at the sixteen fixedpoints (the 2-divisible points) and let K' be the resolution of K. K' is a non-singular K-3 surface,and A' is a double cover of K'.Let H_o be the pullback of the hyperplane section of $\mathbb{P}^3$.(K is naturally a quartic in $\mathbb{P}^3$),and let ΣL be the pullback of H_o on A'.Define M =H+E, where E are the exceptional divisors on A'.E will be a fixed component of M,but will not be a pullback of an effective divisor on K.This is a very typical example of case b) cf. prop.3.2.

We come now to the central proposition in this paper.

Proposition 3.2. Let X be a surface (of general type) such that Φ_K is generically 2:1 onto its image. Then some blow up X' of X is a double cover of a non-singular surface Y. Two cases occur a priori.

I (Babbage) $p_g(Y) = 0$ II K_Y is very ample (in the birational sense)

The examples of I. occur in the following way ,

a) let Y be a non-singular surface with $p_g(Y) = 0$

b) let B be a line bundle such that

i) $|K_Y+B|$ is very ample (in the birational sense)

ii) $|2B|$ contains a smooth divisor C

Given the data (Y,B,C) let X' be a double cover of Y branched along C.

The examples of II. occur in the following way ,

a) let Y be a surface ,with K_Y very ample (birationally)

b) let B be a line bundle such that

i) $|K_Y+B|$ is empty (!)

ii) $|2B|$ contains a smooth curve C.

Given the data (Y,B,C) let X' be a double cover of Y branched along C.

Proof: This is now straightformward from prop.2.9 and prop.3.1. Case a) corresponds to $h^o(K) = 0$, Case b) to $h^o(K+B) = 0$ which gives I and II.

Note: In case I. the conditions i) and ii) on B does not impose any conditions on Y.Thus every Y with $p_g(Y) = 0$ occurs (as a non-singular resolution of the image surface) and almost every normal image of suc surfaces occur. Thus case I gives almost all exceptions to $X^{[1]}$ not being birational to X - the analogy with hyperelliptic curves is striking.Case I was noted by Babbage in [1] (in fact the paper that was the motivation for this article),but he seems to overlook the possibility of case II. In all fairness however II is very hypothetical, the conditions i) and ii) on B here imposes severe restrictions on Y and the curve C. I can e.g prove using R-R that $(K+B)C < 0$,thus by lemma 2.7. C contains at least one rational component with self-intersection -2,and in fact no other kinds of components ,precisely when $h^1(B) = h^1(-B)$.If X is regular then this has to be the case,and I can then show that there has to be at least twenty such components.The logical place to look would consequently be for Kummer-analoges (cf. note after prop.3.1) for quintics in $\mathbb{P}^3$,as Mumford has pointed out to me.I hope to be able later on to settle the existence of case II,I have hitherto been inclined to think of it as a phantom case,but there seem to be no elementary way to rule them out.

The following proposition should be clear.

<u>Proposition 3.3.</u> Let X be a double cover of a ruled surface Y.Then the n-canonical mapping Φ_{nK} factors via Y iff $h^o(nK+(n-1)B) = 0$. (where $2B = C$ is the branch curve).

Note:I conjecture that for $n \geq 2$, Φ_{nK} factors via Y, only if Y is ruled (although there might be exceptions,but which seems exceedingly unlikely).In the last chapter we will discuss the condition of this proposition in detail.

Let me round off this chapter with two propositions due to Beauville ,which ties in with the final loose thread of chapter I.

Proposition 3.4. Let X be a double cover of a non-ruled surface which satisfies condition I or II in prop.3.2. Then $c_1^2 \geq 4p_g-8$.

Proof:Let L be the free part of K+B or K respectively. By corr.2.8 $c_1^2 \geq 2L^2$, so we will be done if we can show $h^o(L) \leq 2 + L^2/2$ .But this is a consequence of Cliffords theorem,as L^2 is special on L for non-ruled surfaces.

Proposition 3.5.(Beauville) Let X be a surface of general type such that $2p_g-4 \leq c_1^2 < 3p_g-7$,then X is a double cover of a ruled surface.

Proof:Clearly $p_g \geq 4$.Assume K composite with a pencil,then $K = F + nC$; F fixed component C irreducible and $4 \leq p_g \leq n+1$.Consider $K^2 = KF + nFC + n^2C^2$ all three terms positive.As $K^2 < 3p_g-7$, $n \geq 4$ $C^2>0$ is impossible.Similarly $K^2 \geq nKC$ shows that $KC \leq 2$, as X is of general type C cannot be rational or elliptic.Thus C genus two,hence hyperelliptic,and hence we get a double cover onto a ruled surface. Now assume that X maps onto a sur face Y.and that the mapping is of degree m.By prop.1.1 and 1.2 m=2 is the only possibilty,and if m=2 ,prop.3.2. and 3.4. conclude that Y is ruled.

This proposition furnishes one motivation for studying double covers of ruled surfaces in more detail,which is the theme of the next chapter.

IV Double coverings of Ruled Surfaces

We are now going to indicate how to get an overall view of all double coverings of ruled surfaces.Because of the Neron-Severi groups simplicity in the ruled case,it lends itself to a very explici analysis.

To put things in perspective we note.

<u>Proposition 4.1</u> Any surface (of general type) which has a hyperelliptic fibration is a double cover of a ruled surface.With the exception of double covers of $\mathbb{P}^2$ with non-singular branch loci,the con verse is true.

Note:The canonical map may have fixed components and may hence not in duce the involution,but will of course always factor through it.(3.2) The exceptions mentioned above,will of course always have a pencil of hyperelliptic curves with two basepoints.

<u>Proposition 4.2.</u> Any surface (of general type) which has a genus two fibration,is a double cover of a ruled surface branched along a sexti section,and conversely such a surface has a genus two fibration.

Note:For such surfaces,clearly the bicanonical map factors through th double covering,but the converse is not true as we shall presently se

Let π:X→Y be a double cover branched at C (=2B),and let p:Y be a morphism onto a minimal model,and let X' be the normalization of X' is a double cover of Y' branched along a -in general singular - br curve C'.Via p there is a one to one correspondence between C and C' proposition 2.3. So the study of double covers reduces to a study of branch curves and their singularities on minimal ruled surfaces. The choice of p is not unique,a natural choice would be to choose p such that it prepares C, see [11] for a definition.

Let us recall the following facts about the Neron-Severi group for minimal ruled surfaces. (see e.g [11])

Let N be the minimal positive selfintersection for sections and let b, be such a section, and let f be a fiber. Now b, f generate the group (integrally), and the intersection matrix is given by $b^2=N, bf=1$ and $f^2=0$. The canonical divisor K is given by $-2b + (N+2q-2)f$, where q = genus of the base curve (= irregularity of the surface) . The most interesting cases happen for rational surfaces, to which we will restrict ourselves from now on. Now the non-negative number N is a complete invariant. We can also be more precise e.g. any divisor $D = mb+nf$, is effective iff $m \geq 0$ and $n \geq -mN$. It can be chosen irreducible (and smooth) iff $m>0$ $n \geq 0$ or $m=0$ $n=1$ or $m=1$ $n=-N$ (the last being the unique section with negative selfintersection $(-N)$) We can now state the salient features of prepardness. C being prepared means (see [11]) that if $C' = 2mb + 2nf$ $m_p(C') \leq m$, and if $N=0$ $n \geq m$.

Now a double cover of a rational surface will be described by a four-tuplet (N,m,n,Q), where everything but the last symbol has been defined. Q will describe the singularities of the branch curve up to inessential ones. (cf. prop.2.5), thus $Q = 0$, means that the branch curve has no infinitely close triple points.

<u>Proposition 4.3.</u> With a "finite" number of exceptions the double cover corresponding to (N,m,n,0) will have the following invariants.

$c_1^2 = 2m(m-2)N + 4(m-2)(n-2)$

$\chi = m(m-1)N/2 + (m-1)(n-1) + 1$

The exceptions being (1,m,0,0) or (2,m,-1,0) where we have resp.

$c_1^2 = 2(m-3)^2$, $\chi = (m-1)(m-2)/2 + 1$ and

$c_1^2 = 4(m-2)(m-3) + 1$, $\chi = (m-1)(m-2) + 1$.

Proof: Clear from the intersection matrix and prop.2.1.

The first series of exceptions occur, because the minimal section in I (for standard notation consult [11]) is exceptional and disjoint from the branch curve, hence the double cover will have two exceptional divisors. (We are really here dealing with double covers of $\mathbb{P}^2$ branched along a non-singular curve.) The second series of exceptions occur, because the minimal section is a component of the branch locus, with self-intersection -2, and hence gives arise to one exceptional divisor upstairs. Cf prop.2.6.)

Note: I believe that all these surfaces are regular.

<u>Proposition 4.4.</u> Throwing away the exceptional cases- in the proposition above- we have the following relationships for fixed m.

$$(m-1)(c_1^2 + 8(m-2)) = 4(m-2)(\chi + (m-2)) .$$

That is, the chern-invariants lie on lines with slopes starting at 2, and asymptotically approaching 4.

Note: When m=3, we get the boundary of the sector $c_1^2 \geq 2\chi - 6$. These surfaces have been studied extensively by Horikawa see [5].

Perhaps somewhat of a curiosity is the fact that all these lines intersect in lattice points !! $((m-1)(m'-1)+1, 4(m-2)(m'-2)) = (\chi, c_1^2)$, but these points only represent surfaces for one of the lines (an exception being m=3, m'=4 with (2,3,1,0), (0,3,4,0) and (1,4,1,0) -due to certain diophantine estimates. Thus there is essentially no overlaps between the different lines.

In order to get the entire collection of double covers of rational surfaces, we have to specialize the ones above i.e imposing essential singularities Q on the branch loci.

Perhaps at this point it might be appropriate to indulge in a little digression. Let us say that a surface Y is a specialization of a surface X, iff we can find a family $\Xi \to \Delta$, with the general fiber birational to X, and Y birational to a component of the special fiber, or stronger birational to the entire special fiber. We thus have a natural partial ordering of all surfaces with respect to specialization. (cf.e.g [11])

Double coverings are particular amendable to such investigations, because their degenerations can be reduced to that of their branch loci (and base surfaces). I hope to return to these questions more systematically in the future.

Thus a surface X corresponding to (N,m,n,Q) is a specialization of a "standard" surface corresponding to (N,m,n,0). Imposing Q will in general lower both c_1^2 and χ ,so we can define a "specialization vector" $-(a,b)_Q$,where ,

$b = c_1^2(X) - c_1^2(X,Q)$, $a = \chi(X) - \chi(X,Q)$

In general the specialization vector will only depend on Q ("abstractly") (thus making sense of the notation), but will have to be modified if the singularities are put in some special position. It is fairly straight forward to compute the specialization vector for various singularities. We can e.g establish the following list.

Q	self explanatory notation	specialization
infinitely close triple point	$3\to 3$	$-(1,1)$
ordinary four-tuple point	4	$-(1,2)$
ordinary 2k-tuple point	2k	$-(k(k-1)/2,2(k-1)^2)$
- " - 2k+1 tuple point	2k+1	- " -
	$6\to 6$	$-(6,16)$ etc.

An interesting phenomena happens for $6\to 6$. Resolving the singularity of the base locus by two blow ups we get $K+B = K'+B' -2e -2e'$ $e'\to e$

Now $|K+B| = \{ S\varepsilon|K+B'|$ having a tacnode at e with tangent direction e'$\}$

there are five conditions for this ,thus $p_g(X,Q) = p_g(X) - 5$,but $\chi(X,Q) = \chi(X) - 6$, hence $q(X,Q) = q(X) + 1$,i.e the irregularity is forced up,by the imposition of the singularity. It would be interesti to deal with these questions more systematically.

Now let us turn to the behaviour of the multi canonical map of the surfaces defined above.

Proposition 4.5, For surfaces corresponding to $(N,m,n,0)$ $|2K+B| = \emptyset$ i.e Φ_{2K} factors through the double cover ,in the following cases.

i) $m = 3$. (Surfaces with genus two fibrations)

ii) $m = 4$ $N = 1$, $n = 0,1$; $N = 2$, $n = -1$

iii) $m \neq 5$ $N = 1$, $n = 0$.

$|3K+2B| = \emptyset$ i.e Φ_{3K} factors through the double cover in these cases. $N = 1$, $m = 4$, $n = 0$ and $N = 2$, $m = 3$, $n = -1$

$|4K+3B| = \emptyset$ i.e Φ_{4K} factors through the double cover in this case $N = 2$,$m = 3$,$n = -1$

The outstanding cases have the following invariants

(1,4,0,0)	$\chi = 4$	$c_1^2 = 3$	$p_g = 3$
(1,4,1,0) (1,5,0,0)	$\chi = 7$	$c_1^2 = 8$	$p_g = 6$
(2,4,-1,0)	$\chi = 7$	$c_1^2 = 9$	$p_g = 6$
(2,3,-1,0)	$\chi = 3$	$c_1^2 = 1$	$p_g = 2$

If we impose Q, more examples can be produced in the bicanonical case

Proof: $2K+B = (m-4)b + (n+2(N-2))f$,$3K+2B = (2m-6)b + (2n+3(N-2))f$ and $4K+3B = (3m-8)b + (3n+4(N-2))f$.

The following proposition will settle the cases n=3,4 definitely.(cf.the list given in the introduction)

Proposition 4.6. Let X be a surface with $c_1^2 = 1$, $p_g=2$,then X correspond to (2,3,-1,0) i.e a double cover of F_2 branched along 5b +(b-2f)

Let X be a surface with $c_1^2 = 2$, $p_g=3$,then X corres

ponds to (1,4,0,0) i.e a double cover of $\mathbb{P}^2$ branched along an octic.

Proof: Assume $c_1^2=1, p_g=2$. The canonical map gives $X\to\mathbb{P}^1$, thus $K=F+C$, where F is a fixed component and C an irreducible curve, $C^2\geq 0$, $(C\neq 0)$ Write $K^2 = KF + CF + C^2$, all three terms being positive. As $KF+F^2 = 2F^2 + CF$, CF is even, now $K^2=1$, hence $CF=0$. Thus $KF=F^2$, if $KF=1$ then $F^2=1$ and $C=0$ by Hodge Index -contradiction. Hence neccessarily $C^2=1$ and $F=0$ (by Hodge Index). Thus K has no fixed component, $K=C$, $KC+C^2=2$, thus C is hyperelliptic (genus two). If we blow up the base point we get a double cover onto a ruled surface, as we have a transversal exceptional divisor we can choose F_2 as our base surface- the rest is clear.

Assume $c_1^2=2, p_g=3$. First we prove that Φ_K cannot be composite with a pencil. Assume so i.e $K = F + nC$, where $n\geq 2$ and C irreducible. Once again all terms in $K^2 = KF + nCF + n^2C^2$ are positive. As $n^2\geq 4$ we conclude $C^2=0$. Thus $CF = KC+C^2$ is even, as $K^2=2, n\geq 2$ we conclude $CF=0$. Thus $KC+C^2=0$ i.e C elliptic, which is a contradiction as X is of general type

Thus Φ_K maps onto $\mathbb{P}^2$ with degree $m\geq 2$ (X is not rational), let $K = F + H$, $K^2 = KF + FH + H^2$, all three terms positive $H^2=m$. Thus $m = 2$ $(K^2=2)$, $KF=FH=0$. Choose a generic line L in $\mathbb{P}^2$, we get a 2:1 map $H\to\mathbb{P}^1$, as $KH + H^2 = 4$, this map has to have eight branchpoints.

Finally as $KF = F^2+FH$, $KF, FH = 0$ we conclude $F^2=0$, thus $F=o$ by H.I.

The behaviour of the multicanonical maps for those surfaces are described below.

<u>Proposition 4.7.</u> Let X be a surface with $c_1^2=1$, $p_g=2$ then Φ_K is composite with a pencil. Φ_{2K} maps X 2:1 onto a quadric cone in $\mathbb{P}^3$ Φ_{3K} maps X 2:1 onto a nonsingular model of F_2 in $\mathbb{P}^5$. Φ_{4K} maps X 2:1 onto the quadric cone embedded by quadric sections into $\mathbb{P}^8$.

Let X be a surface with $c_1^2=2, p_g=3$ then Φ_{nK} maps X 2:1 onto $\mathbb{P}^2$ embedded in $\mathbb{P}^3, \mathbb{P}^5$ and $\mathbb{P}^9$ by the Veronese maps for n = 1,2 and 3.

Proof:Let us just look at a typical case. In the first case for n=3,w have $3K+3B = b + f + (b-2f)$ the last term being a fixed component. Clearly $b+f$ is very ample and $h^0(b+f) = 6$. As $(b+f)(b-2f) = 1$, the fixed component is not contracted by the free part.

Proposition 4.8. Surfaces with $p_g=2,c_1^2=1$ are specializations of surfaces with $p_g=3,c_1^2=2$. The specialization is given by letting the octic branch locus acquire an infinitely close triple points.

Proof:We only have to show that there are octics with one infinitely close triple point and no other -essential- singularities.But choose e.g three quadrics mutually tangent at a given point,and a generic quadric not passing through that point.It is straightforward to compute the invariants of the resulting double cover cf. the list after prop.4.4)

I think I can prove the following (rigorously).

Conjecture:Let X be a surface with $c_1^2=1,p_g=2$,and assume that $\Xi\to\Delta$ is a family of surfaces,the general fiber birational to X.If Y is a component of general type in the special fiber than Y is birational to X

This would give examples of extremal surfaces of general type,and it might be possible to characterize them.

Note:It is clear that if any additional singularities are imposed on an octic worse than a single infinitely close triple point,then the resulting double cover is not of general type anymore.

Let me conclude this chapter by discussing an example by Campadelli,using our techniques,and illustrating the effect of a complicated Q.

Proposition 4.9. Consider a branch locus of three conics pairwise bitangent at a total of six points,and a quartic touching the conics at all the six points.(Such a situation exists).This is a curve of degree ten with six infinitely close triple points.The corresponding double covering has the following properties.

a) $c_1^2 = 2$, $\chi = 1$ and $p_g=0$

b) Φ_{2K} factors via $\mathbb{P}^2$

c) Φ_{3K} gives a birational embedding.

Proof:We first observe by the very construction of the curve,that the six essentially singular points cannot all lie on a conic. Resolve them and we obtain $K+B = 2H -e_1'-..-e_6' = 2H - e_1 \ldots -e_6 +(e_1-e_1')+\ldots(e_6-e_6')$ the six last terms being fixed components.By our initial remark this shows that $p_g = 0$ (hence $\chi = 1$). Now $2(K+B)^2 = -4$, but we can easily see six exceptional divisors in the double cover (coming from the e_i-e_i' ,hence $c_1^2 \geq 2$. Now $2K+B = -H + e_1 + ..e_6 = \emptyset$,and $2K+2B = 4H - 2e_1' - \ldots -2e_6' = 4H - e_1 - \ldots -e_6 - e_1' - \ldots -e_6' +(e_1-e_1')+..(e_6-e_6')$ This imposes 12 independant conditions.(should really be elaborated on) thus $P_2 = 3$ (which shows $c_1^2 = 2$).This settles a) and b) finally to prove c) we need only show that $3K+2B = H +(e_1-e_1')+ \ ..(e_6-e_6') \neq \emptyset$.

Consult [12] for a more exhaustive treatment.

V Multiple Coverings

The natural extension is to consider higher degree coverings and ask when the multicanonical - in particular the canonical maps are m:1 $m>2$. This is a phenomena which occurs ,I will give some examples below,but I suspect it is very rare,hope fully it should be possible to classify them all,however there are technical difficulties mainly of two kinds. To begin with , the proposition 2.9 does not generalize without (apparently at least) serious modifications,we can still find a finite map on the top,but we can no longer assume X' to be non-singular.Secondly even if we consider a finite map between two non-singular surfaces we do not any longer have the same simple relationship between the structure sheaves.Restricting ourselves to cyclic coverings or more generally to abelian,we can work with approximately the same explicitness,but I suspect that the really interesting examples occur for non-Galois coverings, in particular it would be very interesting to elucidate the 3:1 case.

We have sofar the following examples ;

Ex. 5.1. Let $X = \mathrm{Spec}\,\mathcal{O}_{\mathbb{P}^2} \oplus \mathcal{O}_{\mathbb{P}^2}(-2) \oplus \mathcal{O}_{\mathbb{P}^2}(-4)$ with a ringstructure given by an embedding of $\mathcal{O}_{\mathbb{P}^2}(-6)$ into $\mathcal{O}_{\mathbb{P}^2}$. This gives a surface with Φ_K is 3:1 and furthermore is a cyclic map. Are there any more cyclic examples ? The invariants of this surface are $K^2=3$, $\chi=4$ and $p_g = 3$. This example is just a special case of the below.

<u>Proposition 5.2.</u> Let X be any surface with $K^2=3, p_g=3$ then Φ_K has either one or no base points, in the first case Φ_K gives a 2:1 map onto its image and in the second case a 3:1 map onto its image, in both cases the image is $\mathbb{P}^2$.

Proof: We keep the same notation as in the second part of the proof of prop.4.6. That proof shows directly that Φ_K cannot be composite with a pencil, and that $m\leq 3$. Now let us assume m=2, then KF+HF=1. Observe that $HF+F^2=KF$ thus HF is even ($HF = KF+F^2 -2F^2$) hence 0. We hence get $F^2=KF=1$,but as $H^2>0$ this contradicts Hodge Index theorem. Clearly the case m=1 is absurd as X non-rational. This shows that m=3 and the last remark shows that it can be at most one base point. In the case of one basepoint we clearly have a double cover of $\mathbb{P}^2$. (An example can be given by a double covering of $\mathbb{P}^2$ branched along the unioun of ten lines meeting four and four at three points all lying on the "tenth line" - consult prop.4.3. for the proper invariants.) If Φ_K does not have any basepoints then it gives a 3:1 map onto $\mathbb{P}^2$. If this is a cyclic covering the branchlocus will be a sextic -this is our previous example, in the non-cyclic and hence non-Galois case the branchlocus will be a curve of degree 12. It would be instructive to give a concrete example of such a surface,

<u>Ex.5.3.</u> Let F and G be two forms in three variables of degree 4 and 6 respectively. Let Z denote the fiber co-ordinate of the line bundle 2H on $\mathbb{P}^2$, where H is the hyperplane bundle, consider the equation $Z^3+FZ+G=0$ in the completion of the linebundle mentioned before, cf construction before Prop.2.1. This defines a 3:1 covering onto $\mathbb{P}^2$. The branchlocus is given by $4F^3-27G^2$, and the above surface is non-singular exactly when the only singularities of the branchlocus are the 24

"ordinary" cusps.(i.e at the 24 points of intersection between F=0 and G=0,which are required to intersect transversally) Our example 5.1 is a degenerate case occuring when F=0. I believe that this exa le ties in with the one given by Babbage in [1]. It should also be noted that prop.5.2 is treated in detail in [6].

Ex 5.4. Let C and C' be two hyperelliptic curves and let X=CxC'.Then Φ_K gives a 4:1 map onto its image $\mathbb{P}^1 x \mathbb{P}^1$.This is a Galois covering corresponding to $Z_2 x Z_2$.Of course this is an obvious example, a sligh ly less trivial example is given once again by repeated double cover ings of $\mathbb{P}^1 x \mathbb{P}^1$.This time we take as our first branchlocus a curve of degree (2,2n) and for the second covering a branchcurve which is a pullback of a curve of degree (2m,2).(we also assume $m,n \geq 3$).Using prop.3.1 it is easily seen that Φ_K enjoys the same properties as in the trivial example. The above also yields information about the bi canonical map,simply let C and C' be of genus 2 in our first example or m=n=3 in our second version,we will then get cases for which Φ_{2K} is 4:1.

Prop.5.5. deg $\Phi_{2K} \leq 8$. Equality can only occur if $\chi=1$, $K^2=2$.

Proof: Clearly (deg Φ_{2K})(deg ImΦ_{2K}) $\leq 4K^2$ (with equality iff $|2K|$ ha no base points).Now $h^o(2K) = K^2 + \chi$ (follows from Riemann-Roch and t vanishing of $h^1(2K)$) As ImΦ_{2K} is a surface contained in no hyper-plane we get deg Im$\Phi_{2K} \geq K^2 + \chi - 2$.From which the bounds follows.

Remark:The method naturally yields much more precise statements,whic we do not need for the moment.

Ex.5.6. Remarkably the upper bound in 5.5 is achieved. Consider the Campadelli surface described in prop.4.9.One can show (consult [12]) that Φ_{2K} is 8:1 onto $\mathbb{P}^2$. Peters has also informed me (private corres pondence) that the same phenomena occurs for the Burniat surfaces with $K^2=2$.(see also [12]).Both these examples have $p_g=0$.I have learned from Catanese that he has been studying a certain surface with $p_g=q=1$ and $K^2=2$,for which the bicanonical map enjoys the same property. These examples suggest that there should be surfaces for which Φ_K is of quite a high order.We certainly should be able to do better than 4:1,but how high ?

Proposition 5.7. $\deg \Phi_K \leq 36$. Equality can only occur if $\chi=4, K^2 = 36$.

Proof: As before $(\deg \Phi_K)(\deg \operatorname{Im}\Phi_K) \leq K^2$ .Now $\deg \operatorname{Im}\Phi_K \geq p_g-2 \geq \chi-3$. Now apply the highly non-trivial fact $K^2 \leq 9\chi$ (seeIII in chapter I) and we are done.(as in prop.5.5. we do not explicate the various sharpenings that can easily be gotten from the method)

Ex.5.8. Let $X \xrightarrow{\pi} Y$ be a double covering of the Campadelli surface branched along $2K_Y$.($|2K_Y|$ has no base points ([12]) so we can find a smooth representative). Applying prop.3.1. we find that Φ_{K_X} factors via Y and in fact $\Phi_{K_X} = \pi\Phi_{K_Y}$. Hence Φ_K is 16:1 (onto $\mathbb{P}^2$) for our surface X (!). Note that it was crucial that $p_g(Y)=0$.The same construction can also be applied to the examples of Burniat with $K^2=2$.

Can higher degrees occur? In view of prop.5.5. more ingenious constructions are called for.

Finally let me conclude by briefly remarking on the higher dimensional cases.The techniques of double coverings lift without essential modifications to all dimensions,and it would be fruitful to consider double coverings say of rational varieties to get examples of ΦnK not being birational for high n.Without trying to hard I have gotten examples of three-folds where n=7 is necessary,this could be pushed.

Bibliography

[1] Babbage: Multiple canonical surfaces ; Proc.Cambridge Philos.Soc. vol XXX(1934

[2] Bombieri Canonical models of surfaces of general type; Inst.Hautes Etudes Sci.Publ.Math.42(197

Canonical models of surfaces; Manifolds,Proc.Int.Conf.Manifolds and related topics in topology,Tokyo(1973)

[3] Godeaux Les surfaces algebrique non-rationelles des genres arithmetique et geometrique nuls. Actualites Sci.Indust. 123 Exp.Geo.(193

[4] Horikawa On deformations of Quintic surfaces Invent.Math. 31 (1975)

[5] Algebraic surfaces of general type with small c_1^2 I; Ann.of Math. II ser 104 (1976)

[6] Algebraic surfaces of general type with small c_1^2 II; Invent.Math. 37 (1976)

[7] Kodaira Pluricanonical systems on algebraic surfaces of general type; J.Math.Soc.Japan 20 (1968)

[8] Moishezon Surfaces of general type in Shafarevic et al. Algebraic Surfaces; Proc.Steklov.Inst.Math. 75 (1965) Am.Math.Soc.Publ. Providence (1967)

[9] Mumford The canonical ring of an algebraic surface; Ann.of Math. 76 (1962)

[10] Lectures on Curves on an Algebraic surface; Ann.of Math. series 59.Princeton (1966)

[11] Persson On degenerations of algebraic surfaces; Mem.Amer.Math.Soc. 189 (1977)

[12] Peters — On two types of surfaces of general type with vanishing geometric genus; Invent.Math.32 (1976)

On certain examples of surfaces with $p_g=0$ due to Burniat; Nagoya Math.J. 66 (1977)

[13] Ueno — Classification theory of Algebraic Varieties and Compact Complex spaces; Lecture Notes in Mathematics Springer Verlag, Berlin 439 (1975)

[14] du Val — On the ambiguity in the specification of a two-sheeted surface by its branch curve; Proc.Cambridge Philos.Soc.Vol XXX(1934)

[15] International Symposium on Algebraic Topology Mexico (1956)

Some formulas for a surface in $\mathbb{P}^3$

Ragni Piene *

Contents

1. Introduction

In a previous article ([P1]) we defined polar classes for singular projective varieties and gave formulas for the polar classes of a hypersurface. The purpose of the present paper is to apply these results and methods to a surface in $\mathbb{P}^3$ in order to obtain some of the classical formulas of Salmon, Cayley, Zeuthen, and Noether.

In particular we obtain all the classical formulas in the case that the surface $X \subseteq \mathbb{P}^3$ has only ordinary singularities (i.e., a double curve, with a finite number of triple points and a finite number of pinch points), or has "almost ordinary" singularities (i.e the "double curve" may be multiple of a higher order).

It is known ([L1],[R1]) that any smooth surface Z can be mapped finitely to a surface $X \subseteq \mathbb{P}^3$, where X has ordinary singularities, by a generic projection to $\mathbb{P}^3$ of an imbedding $X \subseteq \mathbb{P}^N$. In enumerative geometry, however, one is led to consider also surfaces which are not of this type. Firstly, the singular subvarieties of a variety with ordinary singularities do not themselves have ordinary

* Supported by the Norwegian Research Council for Science and the Humanities.

singularities. For example, the double surface of a threefold in $\mathbb{P}^4$ with ordinary singularities has a triple curve (see Appl.1), and the "surface of contact" on the threefold, with respect to a projection to $\mathbb{P}^3$, aquires a cuspidal curve (see Appl. 2).

Secondly, consider the <u>dual surface</u> $\check{X} \subseteq \check{\mathbb{P}}^3$ of $X \subseteq \mathbb{P}^3$, defined as the closure of the set of tangentplanes to X at smooth points. Then it is almost never true that X and $\check{X}$ both have ordinary singularities. This observation was made already by Salmon, who therefore allowed the surface to have both a double and a cuspidal curve in order to be able to apply the formulas also to the dual surface.

We shall consider a surface X with double and cuspidal curves in Section 5, but we restrict ourselves to the case that the normalization of X is smooth (in the next section we shall comment briefly on how the formulas have to be modified in the general case). Except for one relation (IV_c), that we conjecture but are unable to prove, we obtain also a complete set of formulas in this case. An important special case of the situation described in section 5 is the dual surface of a smooth surface (Appl. 3).

Some of the results below (in particular Thm.2 (I),(II),V), Thm.3 (I), Thm.4 (I)) are contained in my doctoral thesis (M.I.T., 1976). I would like to thank Steven Kleiman, my thesis advisor, for having suggested the topic, and for very helpful discussions particularly about the triple point formulas.

2. <u>Review of general results</u> ([P1]).

Let k be an algebraically closed field. By a <u>surface</u> we mean a purely 2-dimensional, reduced scheme, proper over k. An embedded surface $X \subseteq \mathbb{P}^3 = \mathbb{P}^3_k$ has <u>polar classes</u> $[M_1]$ and $[M_2]$ defined as follows. Let $P \in \mathbb{P}^3$ (resp. $L \subseteq \mathbb{P}^3$) be a point (resp. a line), and let T_x denote the tangent plane to X at a smooth point x. Set

$$M_1 = \text{closure of } \{x \in X \mid x \text{ smooth}, \ P \in T_x\}$$

$$M_2 = \{x \in X \mid x \text{ smooth}, \ L \subseteq T_x\}.$$

Then M_1 and M_2 have natural structures as subschemes of X (reduced if char $k = 0$). If P and L are chosen general, they have codimension 1 and 2 respectively, and the rational equivalence classes $[M_1]$ and $[M_2]$ in the Chow group $A.X$ (of cycles on X modulo rational equivalence) are independent of the particular P and L.

In ([P1]) we found formulas for the polar classes of a hypersurface. We shall state the results for the case of a surface, but first we need some notation.

Let $X \subseteq \mathbb{P}^3$ be a surface and let $\pi: \tilde{X} \to X$ denote the blowup of the jacobian ideal of X, i.e., of the 2nd Fitting ideal $F^2(\Omega^1_X)$ of the module of differentials on X (equal to the ideal generated by the partial derivatives of a polynomial defining X in $\mathbb{P}^3$).

Suppose $f : Z \to X$ is a desingularization of X. Then we let $\varphi: \bar{Z} \to Z$ denote the blowup of the ramification ideal of f, i.e., of the 0th Fitting ideal $F^0(\Omega^1_{Z/X})$ of the module of relative diffenrentials. We have shown ([P1], 2.6) that $f \circ \varphi$ factors through π; let $g: \bar{Z} \to \tilde{X}$ denote the induced map.

We shall use Fulton's intersection theory for singular varieties ([F]). Let $A.Y$ and $A^{\bullet}Y$ denote the Chow group and Chow ring of a scheme Y, and $\cap: A^{\bullet}Y \otimes A.Y \to A.Y$ the cap product.

Define the singular subscheme S of X (resp. the ramification subscheme R of Z) by the ideal $F^2(\Omega^1_X)$ (resp. $F^0(\Omega^1_{Z/X})$). The ith Segre covariant class $s_i = s_i(S,X) \in A_iX$ (resp. $r_i = r_i(R,Z) \in A_iZ$) is defined by :

$$s_1 = \pi_*(c_1(J^{-1}) \cap [\tilde{X}]), \quad s_0 = -\pi_*(c_1(J)^2 \cap [\tilde{X}]),$$

$$r_1 = \varphi_*(c_1(I^{-1}) \cap [\bar{Z}]), \quad r_0 = -\varphi_*(c_1(I)^2 \cap [\bar{Z}]).$$

Note that $s_i = 0$ if $i > \dim S$, and $r_i = 0$ if $i > \dim R$. Finally, set $h = c_1(\mathcal{O}_{\mathbb{P}^3}(1))$, and let $H = h \cap [X]$ denote the class of a hyperplane section of X. From ([P1], 2.3 and 2.5 with $r = 2$) we obtain the following results:

<u>Theorem 1</u>: Let μ_0 denote the degree of the surface $X \subseteq \mathbb{P}^3$. The following equalities hold in $A.X$.

(i) $[M_1] = (\mu_0 - 1)H - s_1$

$$[M_2] = (\mu_0-1)^2 h^2 \cap [X] - 2(\mu_0-1)h \cap s_1 - s_0$$

(ii) $[M_1] = f_*(c_1(\Omega^1_Z) \cap [Z]) + 3H - f_* r_1,$

$$[M_2] = f_*(c_1(\Omega^1_Z)^2 \cap [Z]) + 6h \cap f_*(c_1(\Omega^1_Z) \cap [Z])$$

$$+ 9h^2 \cap [X] - 2f_*(c_1(\Omega^1_Z) \cap r_1) - 6h \cap f_* r_1 - r_0.$$

Now we define the <u>rank</u> μ_1 of X as the degree of $[M_1]$, and the <u>class</u> μ_2 of X as the degree of $[M_2]$. When M_2 is reduced (this is always true if char $k = 0$), μ_2 is indeed equal to the class of X as defined classically, i.e., μ_2 is the number of planes through a given line that are tangent to X at smooth points. (If char $k = p$, it might happen that each point of Supp M_2 occurs with multiplicity p^e, $e > 0$.)

The class μ_2 can also be interpreted as the degree of the dual surface. Define the dual variety $\check{X} \subseteq \check{\mathbb{P}}^3$ of X as the closure of the set of points of $\mathbb{P}^3$ corresponding to planes of $\mathbb{P}^3$ that are tangent to X at a smooth point. The rational map $X \dashrightarrow \check{X}$ extend on $\tilde{X}$ to a morphism $\check{\pi}: \tilde{X} \to \check{X}$, called the dual map (or tangent map or Gauss map) of $X \subseteq \mathbb{P}^3$ ([P1], § 3). We observe that $\check{X}$ need not be a surface, and that if it is, there are two cases :

a) $\check{\pi}$ is birational.

In this case, biduality holds (i.e., $\check{\check{X}} = X$) and $\check{\check{\pi}}$ is equal to the blowup of the jacobian ideal of $\check{X}$ ([P1], 3.3). Then $\mu_2 = \deg \check{X}$ holds.

b) $\deg \check{\pi} = p^e$, some $e > 0$, where $p = \operatorname{char} k$.

In this case, $\mu_2 = p^e \cdot \deg \check{X}$ holds.

For certain kinds of ruled surfaces X, $\check{X}$ is a curve, and $\check{X}$ is a finite number of points iff X is a union of planes. Note that we in all cases have

$$\mu_2 = \deg \check{\pi}_* [\tilde{X}]$$

$$\mu_1 = \deg \pi_* (\check{h} \cap [\tilde{X}]),$$

where $\check{h} = c_1(\mathcal{O}_{\check{\mathbb{P}}^3}(1))$.

Remark 1. Let $S_1 \in A_1 X$ denote the codimension 1 part of t cycle defined by $S \subseteq X$, and write

$S_1 = \sum m_i [S_1^i]$, where the $S_1^i \subseteq X$ are reduced. Then

$$s_1 = \sum e_i [S_1^i]$$

holds, where e_i denotes the multiplicity of $F^2(\Omega^1{}_X)$ in the local ring of X at the generic point of S_1^i.

(ii) If $\dim S = 0$, then $s_1 = 0$, and

$$s_0 = \sum_{x \in S} e_x,$$

where e_x = multiplicity of $F^2(\Omega^1_X)$ in $O_{X,x}$([K],(II,50)).

(Similar computations are valid for r_1 and r_0.)

Now recall the definition of the <u>rank</u> of a space curve, and of the <u>class of immersion</u> of a curve on a surface:

Let $C \subseteq \mathbb{P}^3$ be a (reduced, proper) curve. Then its rank is defined as the class (i.e., degree of the 1st polar class) of a generic projection $\bar{C} \subseteq \mathbb{P}^2$ of C. If C has degree r_0, genus g, and if the ramification divisor of the normalization map $C' \to C$ has degree ν, then the rank r_1 of C is given by

$$r_1 = 2r_0 + 2g - 2 - \nu.$$

Note that the rank of C is equal to the number (possibly counted with multiplicities) of lines that are tangent to C at smooth points and intersect a given line.

Suppose C is a curve on the surface $X \subseteq \mathbb{P}^3$. Then we define the class of immersion ρ of C in X to be

$$\rho = \int \check{h} \cap [C^{\pi}],$$

where C^{π} denotes the strict transform of C by $\pi: \tilde{X} \to X$, $\check{h} = c_1(O_{\check{\mathbb{P}}^3}(1))$, and $\cap$ is via $\check{\pi} : \tilde{X} \to \check{\mathbb{P}}^3$ - provided C is not contained in the singular locus of X. In that case, we define $\tilde{C}$ = the 1-dimensional part of $\pi^{-1}C$ that maps finitely onto C, and set

$$\rho = \int \check{h} \cap [\tilde{C}].$$

In the following three sections we shall consider surfaces $X \subseteq \mathbb{P}^3$ with singularities of three types. Common for these surfaces is that they all have <u>smooth</u> normalization. This restriction is mainly made in order to simplify the computations. It should be clear from the proofs, however, how the formulas would have to be modified if one allows for example also isolated singularities (use Remark 1 (ii) locally). (See also [B], p.162.)

3. Surfaces with ordinary singularities.

Let $X \subseteq \mathbb{P}^3$ be a surface, and denote by $f : Z \to X$ its norm lization. We say that X has ordinary singularities if the follow conditions hold.

(0) char $k \neq 2$

(1) Z is smooth

(2) the singular locus of X is a curve D, and D is smooth except for a finite number of triple points which are triple also for X. There are also a finite number of pinch points on X, situated on D; these are the images of the points of ramification of f. If $x \in D$, then

$$\hat{O}_{X,x} \cong \begin{cases} k[\![t_1,t_2,t_3]\!]/(t_1t_2t_3) & \text{if } x \text{ is triple} \\ k[\![t_1,t_2,t_3]\!]/(t_2{}^2-t_1{}^2t_3) & \text{if } x \text{ is a pinch point} \\ k[\![t_1,t_2,t_3]\!]/(t_1t_2) & \text{otherwise} \end{cases}$$

(3) Let $p : X \to \mathbb{P}^2$ be a generic projection. Then the ramification divisor $C \subseteq Z$ of $p \circ f : Z \to \mathbb{P}^2$, defined by $\mathbf{F}^0(\Omega^1_{Z/\mathbb{P}^2}$ is a smooth curve, and $p \circ f|_C : C \to \mathbb{P}^2$ is a generic immersion (i.e., is generically unramified).

(4) Let $\pi : \tilde{X} \to X$ denote the blowup of the jacobian ideal $F^2(\Omega_X{}^1$ The dual map $\check{\pi} : \tilde{X} \to \check{X} \subseteq \check{\mathbb{P}}^3$ is separable.

Remark 2. Given any smooth surface Z and an embedding $Z \subseteq \mathbb{P}^N$, then a generic projection $Z \to \mathbb{P}^3$ of $Z \subseteq \mathbb{P}^N$ (or of a twis of this embedding) satisfies (1), (2), and (3) if char $k \neq 2$ ([R1 Th.1,3; [R2], §11), and also (4) ([SGA7], Exp.XVII, (2.5) and [W], Th

If char k = 2, we may assume (1) and (2) hold, except that the completion of the local ring at a pinch point looks like $k[[t_1, t_1t_2, t_2^2 + t_2^3]]$. The curve C of (3) need not be smooth, and the map $p \circ f|_C : C \to \mathbb{P}^2$ is everywhere ramified ([R1],[R2]).

Note that a surface $X \subseteq \mathbb{P}^3$ might have ordinary singularities without being obtainable as a generic projection of any embedding of its normalization Z.

Assume now that $X \subseteq \mathbb{P}^3$ has ordinary singularities. Let $\underline{C} = \underline{Hom}_{O_X}(f_*O_Z, O_X)$ denote the conductor of Z in X ($\underline{C}$ is the largest sheaf of ideals of O_X which is also a sheaf of ideals of O_Z). From the local description it follows that $\underline{C}$ defines the curve D of double points (or double curve of X), with reduced structure. Set $\Gamma = f^{-1}D$. Then Γ is defined by $\underline{C}$ in O_Z, and the map $f|_\Gamma : \Gamma \to D$ has degree 2.

In addition to the degree μ_0 of X, its rank μ_1, and its class μ_2, we introduce the following numerical characters. Set

ε_0 = degree of D

ε_1 = rank of D

ρ = class of immersion of D in X

t = # triple points of D (= of X)

γ_2 = # pinch points of X.

Finally, consider the curve of contact $C \subseteq Z$, defined in (4), and the map $\bar{p} = (p \circ f)|_C : C \to \mathbb{P}^2$. Define κ to be the degree of the ramification divisor of $\bar{p}$.

We are now going to prove the five fundamental relations ([B], p.159) that exist between the above 9 characters.

<u>Theorem 2</u>: (I) $\mu_1 = \mu_0(\mu_0-1) - 2\varepsilon_0$

(II) $\mu_1(\mu_0-2) = \kappa + \rho$

(III) $\varepsilon_0(\mu_0-2) = \rho + 3t$

(IV) $2\rho - 2\varepsilon_1 = \nu_2$.

(V) $\mu_2 + 2\nu_2 = \mu_1 + \kappa$.

<u>Proof:</u> Our set-up is as follows:

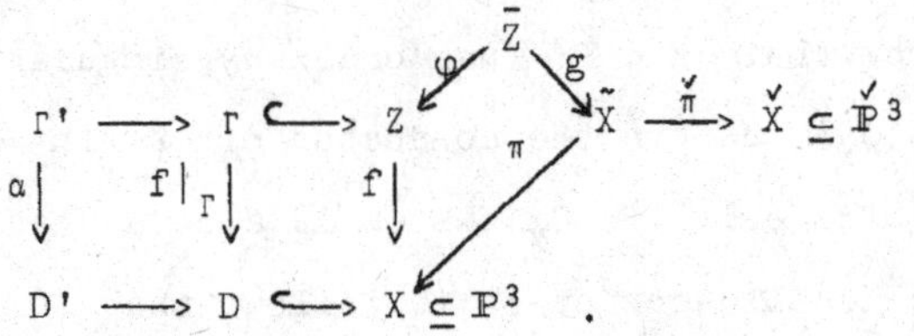

Here π denotes the blowup of the jacobian ideal $F^2(\Omega^1_X)$, φ the blowup of the ramification ideal $F^0(\Omega^1_{Z/X})$, and g is the map whi exists since $F^2(\Omega^1_X)$ becomes invertible on Z ([P1], 2.6). We le D' (resp. Γ') denote the normalization of D (resp. Γ), and $\alpha: \Gamma'$ is the induced map.

<u>Lemma 1:</u> g is isomorphic to the normalization map of $\tilde{X}$

<u>Proof:</u> The ramification ideal $F^0(\Omega^1_{Z/X})$ is equal to maximal ideal of O_Z at a ramification point, and trivial otherwis (this is seen from the local description of X, see also [R3], p.16 Hence φ is the blowup of closed points on the smooth surface Z, that $\bar{Z}$ is also smooth. Thus it suffices to show that g is fini

We shall first show

$$F^2(\Omega^1_X)O_Z = \underline{C}\cdot F^0(\Omega^1_{Z/X}).$$

If $x \in D$ is a double point, $F^2(\Omega^1_X)_x = \underline{C}_x$ holds, because both a equal, in $\hat{O}_{X,x} \cong k[[t_1, t_2, t_3]]/(t_1 t_2)$, to (t_1, t_2). Similarly, if is a triple point, they are both equal to (t_1t_2, t_1t_3, t_2t_3). If x is a pinch point of X, we compute $F^2(\Omega^1_X)$ in

$\hat{O}_{X,x} \cong k[[t_1,t_2,t_3]]/(t_2{}^2-t_1{}^2t_3) \cong k[[t_1,t_1t_2,t_2{}^2]]$ to be $(t_1{}^2,t_1t_2,t_1t_2{}^2)$. Hence $F^2(\Omega^1{}_X)O_Z$ is equal to $(t_1)(t_1,t_2)$ in $\hat{O}_{Z,z} \cong k[[t_1,t_2]]$, which is just $\underline{C}\cdot F^0(\Omega^1{}_{Z/X})$ in $\hat{O}_{Z,z}$ (here $z \in Z$ denotes the point lying over x).

Since $\underline{C}$ is invertible, it follows that $\bar{Z} \to Z$ is also isomorphic to the blowup of $F^2(\Omega^1{}_X)O_Z$, and hence that there is an embedding $\bar{Z} \hookrightarrow Z\times_X\tilde{X}$. Since f is finite, so is $Z\times_X\tilde{X} \to \tilde{X}$, and hence so is g. Q.E.D.

Set $K_Z = c_1(\Omega^1{}_Z) \cap [Z]$ in A_1Z, $K_{\bar{Z}} = c_1(\Omega^1{}_{\bar{Z}}) \cap [\bar{Z}]$ in $A_1\bar{Z}$, $J = F^2(\Omega^1{}_X)O_{\tilde{X}}$, $I = F^0(\Omega^1{}_{Z/X})O_{\bar{Z}}$, $E \subseteq \bar{Z}$ the exceptional divisor of φ (so that, in $A_1\bar{Z}$, $[E] = c_1(I^{-1}) \cap [\bar{Z}]$), and, as in Section 2, we set $h = c_1(O_{\mathbb{P}^3}(1))$, $\check{h} = c_1(O_{\check{\mathbb{P}}^3}(1))$, $H = h \cap [X]$.

By $\int\alpha$ we denote the <u>degree</u> of an element $\alpha \in A_0Y$, and we also write $\int\beta$ for $\int\beta \cap [Y]$, if $\beta \in A^rY$ and the scheme Y has dimension r.

<u>Lemma 2.</u> The number ν_2 of pinch points satisfies

(i) $\nu_2 = -\int c_1(I^{-1})^2$

(ii) $\nu_2 = \int(c_1(\Omega^1{}_Z)^2 - c_2(\Omega^1{}_Z) + 4c_1(\Omega^1_Z) \cap H) + 6\mu_0$.

<u>Proof:</u> (i) follows from the fact that the selfintersection of an exceptional curve (of the first kind) is equal to -1.
(ii) Consider the exact sequence

$$f^*\Omega^1_{\mathbb{P}^3}|_X \to \Omega^1_Z \to \Omega^1_{Z/X} \to 0.$$

Then ν_2 is equal to the degree of the cycle defined by the scheme of zeros of the map $\Lambda^2 f^*\Omega^1{}_{\mathbb{P}^3}|_X \to \Lambda^2\Omega^1{}_Z$, i.e., the scheme defined by $F^0(\Omega^1{}_{Z/X})$. Applying the determinantal formula ([K-L],Cor.11) we obtain (ii). Q.E.D.

<u>Lemma 3:</u> The following equalities of rational equivalence classes of cycles hold.

(i) $\check{h} \cap [\tilde{X}] = (\mu_0-1)(h \cap [\tilde{X}]) - c_1(J^{-1}) \cap [\tilde{X}]$

$\check{h} \cap [\bar{Z}] = \varphi^* K_Z + 3(h \cap [Z]) - [E]$

(ii) $K_{\bar{Z}} = \varphi^* K_Z + [E]$

(iii) $g^* c_1(J^{-1}) \cap [\bar{Z}] = [\varphi^{-1}\Gamma] + [E]$.

<u>Proof:</u> (i) follows from ([P1], 2.3 and 2.4). (ii) is well known - it suffices to consider the exact sequence $0 \to \varphi^*\Omega^1{}_Z \to \Omega^1{}_{\bar{Z}} \to \Omega^1{}_{\bar{Z}/Z} \to 0$, and observe that E is defined by t ideal $F^0(\Omega^1{}_{\bar{Z}/Z})$. (iii) follows from the isomorphism $g^*J \simeq \varphi^*\underline{C}$ (Lemma 1).

Now we shall return to the proof of Theorem 2. First we observe that (I) follows immediately from Theorem 1 (i), since, from what we have seen above, $s_1 = \pi_*(c_1(J^{-1}) \cap [\tilde{X}]) = f_*[\Gamma] = 2[D]$.

(V): Let $C \subseteq Z$ be the ramification curve of a generic p jection $p \circ f : Z \to \mathbb{P}^2$, as in (3), so that C is smooth and $p \circ f|_C : C \to \mathbb{P}^2$ is a generic immersion. Let $\check{\mathbb{P}}^2 \subseteq \check{\mathbb{P}}^3$ be the hyperplane corresponding to the point of projection of p, and let $\bar{C} \subseteq$ be the inverse image of $\check{\mathbb{P}}^2$ via $\check{\pi} \circ g : \bar{Z} \to \check{\mathbb{P}}^3$. Then it is easy t see that $\varphi|_{\bar{C}} : \bar{C} \simeq C$ and $f(C) = M_1$ hold. In particular, the cu $\bar{C}$ has degree μ_1 in $\mathbb{P}^2$.

In order to show (V) we shall compute the class of $\bar{C} \to \mathbb{P}^2$ two ways. First we note that $\check{\pi}$ is separable, by (4). If $\check{X}$ is surface, i.e., if $\mu_2 \neq 0$, we can apply ([P1],4.3), which says that the class of $\bar{C} \to \mathbb{P}^2$ is equal to the degree of $\bar{C} \to \check{\mathbb{P}}^2$, hence to

Let g denote the genus of $\bar{C}$. By the adjunction formula and $\bar{Z}$ are smooth), we obtain

$2g-2 = \overline{C}.(K_{\overline{Z}} + \overline{C}) = \int \check{h} \cap (\varphi^* K_Z + [E] + \check{h} \cap [\overline{Z}]) = \int \check{h} \cap (2\check{h} \cap [\overline{Z}] - 3h \cap [Z] + 2[E])$ (lemma 3) $= 2\mu_2 - 3\mu_1 + 2\int (\check{h} \cap [E]) = 2\mu_2 - 3\mu_1 + 2\nu_2$ (lemma 3, and the projection formula).

Applying one of the Plücker formulas ([P1], 3.9 (II)) we find that the class is equal to

$$2\mu_1 + 2g - 2 - \kappa,$$

hence we get

$$\mu_2 = 2\mu_1 + 2\mu_2 - 3\mu_1 + 2\nu_2 - \kappa.$$

Suppose now $\mu_2 = 0$. Then either $\check{X}$ is a union of points, in which case X is a union of planes, and all the characters occurring in (V) are then 0. Or $\check{X}$ is a curve. Then ([W],Thm.3) $f(C)$ is a union of lines, hence the class of C (w.r.t. $C \to \mathbb{P}^2$) is 0, and the last part of the reasoning is still valid.

(II) We compute

$$\mu_2 = \int \check{h}^2 = \int \check{h} \cap ((\mu_0 - 1)H - c_1(J^{-1}) \cap [\tilde{X}])$$

$$= \mu_1(\mu_0 - 1) - \int \check{h} \cap ([\varphi^{-1}\Gamma] + [E]) \quad \text{(lemma 3 (iii))}$$

$$= \mu_1(\mu_0 - 1) - \rho - 2\nu_2,$$

because ρ is equal to $\int \check{h} \cap$ (strict transform of Γ via φ), by definition, and this strict transform is equal to $[\varphi^{-1}\Gamma] - [E]$. Together with (V), this proves (II).

(IV) The rational map $D \dashrightarrow G = \mathrm{Grass}_2(V)$ which sends a smooth point of D to its tangent line, extends to a morphism $\gamma : D' \to G$, called the first associated map of $D \subseteq \mathbb{P}^3$, and the degree of the curve $D' \to G$ (via the Plücker embedding $G \hookrightarrow \mathbb{P}(\Lambda^2 V)$) is the same as the rank ε_1 of D ([P3], §3).

The normalization maps $\Gamma' \to \Gamma$ and $D' \to D$ are both unramified, since D has only t triple points and Γ has $3t$ (ordinary) double points, as singularities. Therefore $\alpha: \Gamma' \to D'$ is a morph of degree 2 with ν_2 ramification points. We shall now show that the map $\gamma \circ \alpha: \Gamma' \to G$ can be defined in another way, using the fa that at a (general) point of D there are two distinct tangent pla to X, and their intersection is the tangent line to D at the poi

Consider the secant map

$$\check{\mathbb{P}}^3 \times \check{\mathbb{P}}^3 \dashrightarrow G^* = \mathrm{Grass}_2(V^\vee) ,$$

which is defined outside the diagonal and sends a pair of points to the line joining them (or, a pair of planes in $\mathbb{P}^3$ to their line o intersection). Let $b: B \to \check{\mathbb{P}}^3 \times \check{\mathbb{P}}^3$ denote the blowup of the diagona Then ([L],4.2;[K], (V,41-42)) the morphism $B \to G^*$ is defined by 2-quotient $V_B^\vee \to \mathcal{E}$, where $\mathcal{E}$ fits into an exact sequence

$$0 \to \mathfrak{A} \otimes \mathcal{L}_1 \to \mathcal{E} \to \mathcal{L}_2 \to 0$$

with $\mathcal{L}_i = b^* pr_i^* O_{\check{\mathbb{P}}^3}(1)$, and where $\mathfrak{A} = O_B(1)$ denotes the ideal of the exceptional divisor.

Set $\overline{\Gamma}$ = closure of $\Gamma' x_{D'} \Gamma' - \Delta$, where Δ = diagonal. The the projection on the first factor induces a map $\overline{\Gamma} \to \Gamma'$, which is necessarily an isomorphism.

Let $s : \Gamma' \to \Gamma' x_{D'} \Gamma'$ denote the corresponding section (so that $s = (id, i)$, where $i: \Gamma' \to \Gamma'$ corresponds to the involution $\overline{\Gamma}$ induced by the switch of factors on $\Gamma' x_{D'} \Gamma'$). Now the dual map of $X \subseteq \mathbb{P}^3$ is defined on Z, except at the ramification points of Therefore there is a rational map

$$Z \times Z \dashrightarrow \check{\mathbb{P}}^3 \times \check{\mathbb{P}}^3 ,$$

hence also $\Gamma' x_{D'} \Gamma' \dashrightarrow \check{\mathbb{P}}^3 \times \check{\mathbb{P}}^3$, which composed with $s: \Gamma' \to \Gamma' x_{D'} \Gamma'$ gives a morphism (since Γ' is a smooth curve the rational map is everywhere defined) $\Gamma' \to \check{\mathbb{P}}^3 \times \check{\mathbb{P}}^3$. Since the preimage of the diagon in $\check{\mathbb{P}}^3 \times \check{\mathbb{P}}^3$ is equal to $s^{-1}(\Delta)$ and therefore is invertible on Γ' this morphism lifts to a morphism $\sigma: \Gamma' \to B$. Identifying G^* and via the canonical isomorphism, we thus have obtained a factorization

$$\Gamma' \overset{\sigma}{\to} B \to G^* \cong G$$

of $\gamma \circ \alpha$.

Let us compute the degree of $\Gamma' \to G$ (earlier seen to be $2\,\varepsilon_1$) using this factorization. The map is defined by the 2-quotient $V^\vee_{\Gamma'} \to \sigma^*\mathcal{E}$, hence we must compute $\int c_1(\sigma^*\mathcal{E})$. From the exact sequence above we get

$$c_1(\sigma^*\mathcal{E}) = c_1(\sigma^*\mathcal{L}_1) + c_1(\sigma^*\mathcal{L}_2) + c_1(\sigma^*\mathfrak{a}).$$

We let $\mathcal{L}^*$ denote the pullback of $\mathcal{O}_{\check{\mathbb{P}}^3}(1)$ to Γ' via the map induced by the dual map of X ($\Gamma' \to \check{\mathbb{P}}^3$ is a priori a rational map, but extends since Γ' is a smooth curve). Then we observe : $\sigma^*\mathcal{L}_1 = \mathcal{L}^*$, $\sigma^*\mathcal{L}_2 = i^*\mathcal{L}^*$. Since $\sigma^*\mathfrak{a}$ = ideal of $s^{-1}(\Delta)$ in Γ', we get $c_1(\sigma^*\mathfrak{a}) = c_1(F^0(\Omega^1_{\Gamma'/D'}))$, and hence

$$\int c_1(\sigma^*\mathcal{E}) = 2\int c_1(\mathcal{L}^*) - c_1(F^0(\Omega^1_{\Gamma'/D'})^{-1})$$

$$= 2\rho - \nu_2 .$$

<u>Remark 3:</u> The formula just proved, $2\rho - 2\,\varepsilon_1 = \nu_2$, is a version of the <u>correspondence principle</u> ([S-R],p.64). Namely, we are considering a simply infinite system $(H,H';\,l)$ of unordered plane pairs together with a line contained in both. Then: ρ = # of forms such that H contains a given point, ε_1 = # of forms such that l meets a given line, and the "number of coincidences", i.e.,of forms such that H and H' coincide, is equal to the number of pinch points and to $2(\rho - \varepsilon_1)$.

In section 4 we show that our proof of (V) generalizes to the case of a surface with a higher order multiple curve.

(III): Suppose we prove

$$(*) \qquad 2\,\varepsilon_1 + \nu_2 = 2(\mu_0 - 2)\,\varepsilon_0 - 6\,t.$$

Then (III) follows, using (IV).

Let g_a and g denote the arithmetic and geometric genus of

Since Γ has $3t$ ordinary double points, $g_a = g + 3t$. The adjuncti formula (and lemma 3) gives

$$2g_a - 2 = \Gamma.(K_Z + \Gamma) = \int (\mu_0 - 4)h \cap [\Gamma] = 2(\mu_0 - 4)\,\varepsilon_0 .$$

Let $g(D)$ denote the genus of D. The rank ε_1 of D is given b

$$\varepsilon_1 = 2\varepsilon_0 + 2g(D) - 2.$$

Finally, the Riemann-Hurwitz formula gives

$$2g - 2 - \nu_2 = 2(2g(D) - 2),$$

and putting the above together, we obtain (*).

QED

Most of the above formulas are valid also for a surface $X \subseteq \mathbb{P}$ which has ordinary singularities in the following weaker sense.

Assume $X \subseteq \mathbb{P}^3$ satisfies (1) and (2)', where (2)' is the co dition (2) with the following addition :

If char $k = 2$, the completion of the local ring of X at a pinch point is isomorphic to

$$k[[t_1, t_1 t_2, t_2^2 + t_2^3]] .$$

The numerical characters of X are still well defined, except possibly κ. Now we define ν_2 as the degree of the ramification scheme of f (defined by the ideal $F^0(\Omega^1_{Z/X})$), and we note (use th local description, see [R3], p.162) that ν_2 is equal to the number of pinch points except if char $k = 2$, and then ν_2 is equ to twice the number of pinchpoints.

Going through the proof of Theorem 2, one checks that everythi holds for a surface X satisfying (1) and (2)' except possibly formula (V) and the ones depending on it.

Up to now we have been concerned with the "extrinsic" characters of the surface, i.e., such that depend on the given map to projective space. A smooth surface Z has also "intrinsic" invariants. We set

$$c_1{}^2 = \int c_1(\Omega^1{}_Z)^2 ,$$

$$c_2 = \int c_2(\Omega^1{}_Z) ,$$

$$\chi(O_Z) = \sum_{i=0}^{2} (-1)^i \dim_k H^i(Z,O_Z) .$$

Hence $c_1{}^2$ is the selfintersection of a canonical divisor ("canonical grade"), c_2 is the Euler characteristic, and $\chi(O_Z)$ is the Euler-Poincaré characteristic, equal to $p_a + 1$, where p_a is the arithmetic genus of Z.

These three characters are related by Noether's formula

$$12\chi(O_Z) = c_1{}^2 + c_2 .$$

In ([P2]) we gave a proof of this formula along classical lines, using the fact ([R1],Thm.1,3) that any smooth surface Z can be obtained as the normalization of a surface $X \subseteq \mathbb{P}^3$ which satisfies (1) and (2)'. The proof consisted in expressing the three intrinsic invariants in terms of the extrinsic ones, as follows.

<u>Proposition 1</u>.

(i) $$c_1{}^2 = \mu_0(\mu_0-4)^2 - (3\mu_0 -16)\varepsilon_0 + 3t - \nu_2$$

(ii) $$c_2 = \mu_0(\mu_0{}^2-4\mu_0 + 6) - (3\mu_0 -8)\varepsilon_0 + 3t - 2\nu_2$$

(iii) $$\chi(O_Z) - 1 = \binom{\mu_0-1}{3} - \tfrac{1}{2}(\mu_0-4)\varepsilon_0 + \tfrac{1}{2}t - \tfrac{1}{4}\nu_2 .$$

In the proof of (i) and (ii), however, we used Kleiman's triple point formula ([K],(I,39)) :

<u>Proposition 2</u>. $3t = (\Gamma^2) - \mu_0\varepsilon_0 + \nu_2$.

We shall now show that this formula is a corollary of (III) of Theorem 2.

<u>Proof of Prop. 2</u>. Compute (using lemma 3)

$$\rho + (\Gamma^2) = \int \check{h} \cap ([\varphi^{-1}\Gamma] - [E]) + (\Gamma^2)$$

$$= (\mu_0 - 1)\int h \cap [\Gamma] - (E, \varphi^{-1}\Gamma) - \int \check{h} \cap [E]$$

$$= 2(\mu_0 - 1)\varepsilon_0 - \nu_2 .$$

We also have the following classical formulas ([B],p.197):

<u>Proposition 3</u>. (i) $c_1{}^2 = \mu_2 - 6\mu_1 + 9\mu_0 + \nu_2$

(ii) $c_2 = \mu_2 - 2\mu_1 + 3\mu_0$.

<u>Proof.</u> This is a corollary of Proposition 1 and Theorem 2. simpler proof is obtained directly, by using lemma 3 (i) to compute and then lemma 2 (ii) together with (i) to compute (ii).

Q.E

For other formulas that can be deduced from the ones above, we refer to the full discussion given in ([B], Ch. IV and V). Examples are also given there, in particular fifteen of Noether's examples.

We shall close this section with an application of another resu of ([P1]) concerning the dual surface. In particular, it shows tha the dual surface of a surface with ordinary singularities almost alw has a cuspidal curve.

<u>Proposition 4.</u> Let $X \subseteq \mathbb{P}^3$ be a surface which satisfies (1), (2)', and (4), and assume also that $\check{X}$ is a surface. Then the map $\check{\pi} \circ g \colon \overline{Z} \to \check{\mathbb{P}}^3$ ramifies along a curve $\check{R}$, whose degree with respect t

$\mathbb{P}^3$ is

$$4(\mu_0(\mu_0 - 2) - 2\varepsilon_0) ,$$

and with respect to $\check{\mathbb{P}}^3$ is

$$4\mu_0(\mu_0-1)(\mu_0-2) - 24(\mu_0-2)\varepsilon_0 + 12\varepsilon_1 + 48t$$

([B], p.162).

<u>Proof.</u> It follows from ([P1],3.4) that

$$[\check{R}]-2[E] = \int 4(\check{h}-h) \cap [\bar{Z}]$$

holds,

$$\deg_{\mathbb{P}^3}[\check{R}] = \int h \cap [\check{R}] = 4(\mu_1-\mu_0)$$

and

$$\deg_{\check{\mathbb{P}}^3}[\check{R}] = \int \check{h} \cap [\check{R}] = 4(\mu_2-\mu_1) + 2\int \check{h} \cap [E] .$$

Lemma 3 implies $\int \check{h} \cap [E] = \nu_2$, and the proposition now follows from Theorem 2, (IV), (I), and (III).

Q.E.D.

Finally, let us just remark that each of the exceptional lines of $\bar{Z}$ (with respect to the blowup φ) is mapped to a line in $\check{\mathbb{P}}^3$. In fact, let $E_1 \subseteq \bar{Z}$ be such a line. Then

$$\deg \check{\pi}_*[E_1] = \int \check{h} \cap [E_1] = -(E,E_1) = 1.$$

Hence the dual surface $\check{X}$ contains at least ν_2 lines.

<u>Example 1</u>. The Steiner surface.

Let $V \subseteq \mathbb{P}^5$ denote the Veronese surface (i.e., V is isomorphic to $\mathbb{P}^2$ via the embedding $\mathbb{P}^2 \to \mathbb{P}^5$ by conics), and let $f : V \to \mathbb{P}^3$ be a generic projection. Then the image $S = f(V)$ has ordinary singularities; it is the socalled Steiner surface.

The intrinsic invariants of S are equal to those of $\mathbb{P}^2$, hence

$$c_1{}^2 = 9, \quad c_2 = 3, \quad p_a = 0.$$

Since V has degree 4 in $\mathbb{P}^5$, so does S in $\mathbb{P}^3$. We can compute the number of pinchpoints (Lemma 2 (ii))

$$\nu_2 = c_1{}^2 - c_2 + 4 \int h \cap K_{\mathbb{P}^2} + 6\,\mu_0 = 6 .$$

Then the other characters of S are deduced from Thm. 2 and Prop. 1

$$\mu_1 = 6, \quad \mu_2 = 3, \quad \varepsilon_0 = 3, \quad \kappa = 9, \quad \rho = 3, \; \varepsilon_1 = 0, \quad t =$$

Moreover, the (reducible) double curve $D \subseteq S$ and its preimage $\Gamma \subseteq V$ have arithmetic genus 0 and 1 respectively, while both curves have geometric genus equal to -2.

Geometrically, this shows that D is equal to the union of thr (not co-planar) lines intersecting in one point, which is of course well known ([S-R], p.134).

Example 2. Enriques surface (char $k \neq 2$).

Let Z be a (smooth) Enriques surface, i.e., Z satisfies

$$K_Z \neq 0 , \quad 2K_Z = 0 , \quad \dim H^1(Z,O_Z) = 0 .$$

Artin showed ([A], Thm.3.1.1) that Z can be mapped birationally onto a surface $X \subseteq \mathbb{P}^3$ of degree 6, and that in general X has the edges of a tetrahedron as singular locus.

Assuming X has ordinary singularities then the statement abou the nature of the singular locus holds, by the formulas of this section:

Since K_Z is numerically trivial, and since $K_Z \neq 0$ implies $p_g = \dim H^0(Z,K_Z) = 0$, the intrinsic characters are :

$$c_1{}^2 = 0, \quad c_2 = 12, \quad p_a = 0 .$$

As in Ex. 1 we compute ν_2 :

$$\nu_2 = c_1{}^2 - c_2 + 4\int h \cap K_Z + 6\mu_0 = 24 .$$

The remaining characters are :

$$\mu_1 = 18, \quad \mu_2 = 30, \quad \varepsilon_0 = 6, \quad \kappa = 60, \quad \rho = 12, \quad \varepsilon_1 = 0, \; t = 4.$$

Since $\varepsilon_1 = 0$, the double curve $D \subseteq S$ is a union of lines. Moreover, we find that D has arithmetic genus 3 and geometric genus -5, while the genera of $\Gamma = f^{-1}(D) \subseteq Z$ are 13 and 1 respectively.

4. Surfaces with multiple curves.

In this section we shall generalize the results of the preceding one to the case that the surface $X \subseteq \mathbb{P}^3$ is allowed to have, instead of a double curve, a j-multiple curve, for some integer j. This is a special case of a type of surface studied by Noether ([N]). Here we shall assume for simplicity that X has no isolated singularities, and that all the components of the singular curve have the same multiplicity j (but see Remark 4).

Let $X \subseteq \mathbb{P}^3$ be a surface satisfying (0), (1), (3), (4) of the preceding section, and also

(2)j : The singular locus of X consists of a curve D of points of multiplicity j. At a general point on this curve the completion of the local ring of X is isomorphic to

$$k[t_1, t_2, t_3]/(l_1 \, l_2 \cdots l_j) \quad ,$$

where the l_i's are linear and such that l_i and l_j are linearly independent, for $i \neq j$.

There are a finite number of pinch points on X, situated on D; these are the images of the ramification points of f.

We set $D \subseteq X$ equal to the singular curve of X with reduced structure, and set $\Gamma = f^{-1}D \subseteq Z$. The degree μ_0, rank μ_1, and cla μ_2 are defined as before, and so are the degree ε_0, rank ε_1, clas of immersion ρ of D. Again ν_2 denotes the number of pinchpoint (defined as the degree of the ramification cycle of f), κ is the degree of the ramification divisor of the projection $p \cdot f|_C : C \to \mathbb{P}^2$ a curve of contact $C \subseteq Z$ (see (4)).

The character "replacing" the number t of triple points of a surface with ordinary singularities, is "the number of double points of Γ", δ, which we define as the difference $g_a(\Gamma) - g(\Gamma)$ between the arithmetic and geometric genera of Γ.

Theorem 3:

$$(I)\ \mu_1 = \mu_0(\mu_0 - 1) - j(j-1)\varepsilon_0$$

$$(II)\ \mu_1(\mu_0 - 2) = (j-1)\rho + \kappa + (j-2)\nu_2$$

$$(III)\ j\varepsilon_0(\mu_0 - j) = j\rho + 2(j-1)\delta$$

$$(IV)\ 2(j-1)\rho - j(j-1)\varepsilon_1 = \nu_2$$

$$(V)\ \mu_2 + 2\nu_2 = \mu_1 + \kappa .$$

Proof: The set up and notations are as in the proof of Thm and the proof is of course similar (partly identical) to that proof.

First we observe that $\Gamma = f^{-1}D$ is reduced, $\Gamma \to D$ has degree j and ν_2 ramification points. The conductor $\underline{C}$ defines on Z t cycle $(j-1)\Gamma$. Since $g^*J \simeq \varphi^*\underline{C} \cdot I$ holds, we get $s_1 = j(j-1)[D]$, thus (I) is proved (Thm.1 (i)).

Moreover, lemma 2 remains true, and so does lemma 3, provided w replace (iii) by

$$(iii)_j \qquad g^* c_1(J^{-1}) \cap [\overline{Z}] = (j-1)[\varphi^{-1}\Gamma] + [E].$$

The proof of (V) is as before. To prove (II), consider

$$\mu_2 = \int \check{h}^2 = \int \check{h} \cap ((\mu_0 - 1)H - (j-1)[\varphi^{-1}\Gamma] - [E])$$

$$= (\mu_0 - 1)\mu_1 - (j-1) \int \check{h} \cap ([\varphi^{-1}\Gamma] - [E]) - j \int \check{h} \cap [E]$$

$$= \mu_1(\mu_0 - 1) - (j-1)\rho - j\nu_2 \ .$$

Now substitute for μ_2 using (V).

(IV) The proof of Thm.2 (IV) generalizes; we shall keep the no-
tations. Let $n : \Gamma'' \to \overline{\Gamma}$ denote the normalization of the closure $\overline{\Gamma}$ of $\Gamma' \times_{D'} \Gamma' - \Delta$, and let $i: \overline{\Gamma} \to \overline{\Gamma}$ denote the involution, induced by switching the factors of the product. Let $q : \Gamma' \to \check{\mathbb{P}}^3$ denote the morphism induced from the dual map $\check{\pi}: Z \dashrightarrow \check{\mathbb{P}}^3$, and $p: \overline{\Gamma} \to \Gamma'$ the map induced by projecting onto the first factor. As in Thm. 2, the map $\Gamma'' \to \check{\mathbb{P}}^3 \times \check{\mathbb{P}}^3$ lifts to B, and we denote this map by $\sigma: \Gamma'' \to B$. Thus $\Gamma'' \to G^*$ is defined by $\check{V}_{\Gamma''} \to \sigma^*\mathcal{E}$, where $\mathcal{E}$ fits into an exact sequence $0 \to I \otimes \mathcal{L}_1 \to \mathcal{E} \to \mathcal{L}_2 \to 0$, as before. We get:

$$\sigma^*\mathcal{L}_1 = \sigma^* b^* pr_1{}^* \mathcal{O}_{\check{\mathbb{P}}^3}(1) = n^* p^* q^* \mathcal{O}_{\check{\mathbb{P}}^3}(1) \ ,$$

$$\sigma^*\mathcal{L}_2 = \sigma^* b^* pr_2{}^* \mathcal{O}_{\check{\mathbb{P}}^3}(1) = n^* i^* p^* q^* \mathcal{O}_{\check{\mathbb{P}}^3}(1) \ .$$

ince p has degree j-1, we obtain

$$\int c_1(\sigma^*\mathcal{L}_1) = \int c_1(\sigma^*\mathcal{L}_2) = (j-1)\rho \ .$$

ext we observe that the ideal I pulled back to $\Gamma' \times_{D'} \Gamma'$ is equal o the ideal of the diagonal $\Delta \subseteq \Gamma' \times_{D'} \Gamma'$. Therefore $c_1(\sigma^* I^{-1}) = (\Delta . \overline{\Gamma})$, as divisors on $\Gamma' \times \Gamma'$. Moreover, since $= \Gamma' \times_{D'} \Gamma' - \Delta$ and $\Gamma' \times_{D'} \Gamma' = (\alpha \times \alpha)^{-1}(\Delta_{D' \times D'})$, straightforward omputation yields

$$\int c_1(\sigma^* I^{-1}) = - \int \alpha^* c_1(\Omega^1_{D'}) + c_1(\Omega^1_{\Gamma'})$$

$$= \int c_1(F^0(\Omega^1_{\Gamma'/D'})^{-1}) = \nu_2 .$$

Hence the degree of the curve $\Gamma'' \to G^*$ is equal to

$$2(j-1)\rho - \nu_2 .$$

On the other hand, since this map is equal to the composition

$$\Gamma'' \overset{n}{\to} \overline{\Gamma} \overset{p}{\to} \Gamma' \overset{\alpha}{\to} D' \overset{\gamma}{\to} G \cong G^* ;$$

since the rank of D' (equal to the degree of the 1st associated curve γ of D') is equal to ε_1 (by definition), and α and p have degrees j and $j-1$ respectively, the degree of the above curve is also equal to $j(j-1)\varepsilon_1$. This proves (IV).

(III) We shall first prove

$$(*)j \qquad j\varepsilon_1 + \nu_2 = j(\mu_0-2)\varepsilon_0 - 2\delta - (j-2)(\Gamma^2)$$

$$(**)j \qquad \rho = (\mu_0-1)j\varepsilon_0 - (j-1)(\Gamma^2) - \nu_2 .$$

Let g denote the genus of Γ. Then the adjunction formula gives

$$2g - 2 + 2\delta = (\Gamma.K_Z + \Gamma) = (\Gamma.K_Z + (j-1)\Gamma - (j-2)\Gamma) ,$$

$$j\varepsilon_1 - 2j\varepsilon_0 + \nu_2 + 2\delta = (\mu_0-4)j\varepsilon_0 - (j-2)(\Gamma^2),$$

from lemma 3, and from the Riemann-Hurwitz formula applied to $\alpha: \Gamma' \to D'$, as well as the formula for the rank of D'.

To prove (**)j consider

$$\rho = \int \acute{h} \cap ([\varphi^{-1}\Gamma] - [E]) \qquad \text{(by definition)}$$

$$= \int (\mu_0-1)h \cap [\varphi^{-1}\Gamma] - (j-1)(\varphi^{-1}\Gamma^2) + (E^2)$$

(by lemma 3 and the projection formula)

$$= (\mu_0-1)j\varepsilon_0 - (j-1)(\Gamma^2) - \nu_2 .$$

Eliminating (Γ^2) from (*)j and (**)j and using (IV) to elimin[ate] ε_1 , we get (III).

Q.E.D

Corollary 1:

(i) $j\nu_2 = j(j-1)\{2(\mu_0-j)\varepsilon_0-j\varepsilon_1\} - 4(j-1)^2\delta$

(Noether [N], see [B], p.163).

(ii) $\mu_2 = \mu_0(\mu_0-1)^2 - (j-1)\{(3j+1)\mu_0 - 2j(j+1)\}\varepsilon_0$

$+ j^2(j-1)\varepsilon_1 + \frac{2}{j}(j-1)^2(2j+1)\delta$

(Salmon [S], p.308; Noether [N], see [B], p.163).

Proof: (i) follows from (IV) and (III).

(ii) follows from (V), (II), (I) + (III), (i).

From (III) and $(**)_j$ we deduce the following formula for the double points of Γ.

Corollary 2: $2(j-1)\delta = j(j-1)(\Gamma^2) - j(j-1)\mu_0\varepsilon_0 + j\nu_2$.

This formula yields the triple point formula (Prop. 2) when X has ordinary singularities: then $j = 2$ and $\delta = 3t$.

Again, let us look at the intrinsic invariants. Note that the formulas of Prop.3 also hold in this case, so that we can use Thm.3 and Cor. 1 to obtain formulas for $c_1{}^2$ and c_2. Then we use Noether's formula to obtain an expression for $\chi(O_Z) = p_a + 1$.

Proposition 5:

(i) $c_1{}^2 = \mu_0(\mu_0-4)^2 - (j-1)\{(3j-1)\mu_0 - 2j(j+3)\}\varepsilon_0$

$+ j(j-1)^2\varepsilon_1 + \frac{2}{j}(j-1)^2(2j-1)\delta$.

(ii) $c_2 = \mu_0(\mu_0{}^2-4\mu_0+6) - (j-1)\{(3j+1)\mu_0 - 2j(j+2)\}\varepsilon_0$

$+ j^2(j-1)\varepsilon_1 + \frac{2}{j}(j-1)^2(2j+1)\delta$.

(iii) $$p_a = \binom{\mu_0-1}{3} - \frac{1}{6}j(j-1)(3\mu_0-2j-5)\varepsilon_0 + \frac{1}{12}j(j-1)(2j-1)\varepsilon_1 + \frac{2}{3}(j-1)^2\delta$$

(Noether [N], see [B], p.163,p.180).

<u>Remark 4:</u> If X has curves of different multiplicities, t methods above still yield formulas.

As an example, suppose that X, in addition to a double curve degree ε_0, rank ε_1, class of immersion ρ, with ν_2 of the pinch points of X on it, has a <u>triple</u> curve of degree t, rank r_t, clas of immersion ρ_t, and with ν_2^t of the pinchpoints of X on it, th if Γ_t denotes the inverse image on Z of the triple curve, and if we set $i = (\Gamma,\Gamma_t)$, the following formulas are seen to hold (simply go through the proof of Thm. 3 and see where it has to be modified according to the intersections i of Γ with Γ_t).

(I) $$\mu_1 = \mu_0(\mu_0-1) - 2\varepsilon_0 - 6t$$

(II) $$\mu_1(\mu_0-2) = \rho + 2\rho_t + \kappa + \nu_2^t$$

(III) $$\varepsilon_0(\mu_0-2) = \rho + \delta + i$$

(III_t) $$3t(\mu_0-3) = 3\rho_t + 4\delta_t + i$$

(IV) $$2\rho - 2\varepsilon_1 = \nu_2$$

(IV_t) $$4\rho_t - 6r_t = \nu_2^t$$

(V) $$\mu_2 + 2\nu_2 + 2\nu_2^t = \mu_1 + \kappa .$$

Application 1. The double surface of a 3-fold with ordinary singularities in $\mathbb{P}^4$. (See Roth [R].)

Let $V \subseteq \mathbb{P}^4$ be a 3-fold with ordinary singularities ([R]), and let $B \subseteq V$ denote its double surface. Then B has a triple curve. Let $B_0 \subseteq \mathbb{P}^3$ be a generic projection of B - then B_0 has in addition to the triple curve a double curve. With the notations of Remark 4, the class μ_2 of B_0 is given by

$$\mu_2 = \mu_0(\mu_0-1)^2 - \varepsilon_0(3\mu_0-4) - 4t(2\mu_0-3) + \delta + \tfrac{8}{3}\delta_t + \tfrac{5}{3} i - 3\nu_2^t - 2\nu_2 .$$

The terms involving only the double curve (i.e., the 1st, 2nd, 4th, and last term) agree with the corresponding terms as found by Roth ([R], (42)):

$$\mu_2 = \mu_0(\mu_0-1)^2 - \varepsilon_0(3\mu_0-4) - 12t - 15r_t + \delta - 60\xi - \tfrac{13}{2}\nu_2{}^t - 2\nu_2 .$$

(here ξ denotes the number of quadruple points on the triple curve; these points are 6-tuple for B_0 and lie also on the double curve). The other terms, however, do not agree, even when we substitute for r_t and try to interpret i and δ_t in terms of ξ. This sad fact is less surprising when one notices that Roth's formula which corresponds to (IV_t) of Remark 4, reads ([R], (37)):

$$2\rho_t - 4r_t = \nu_2{}^t .$$

Roth obtains it by applying Chasles' correspondence theorem: If the points on a line are in (α,β) correspondence, then their number of united points is equal to $\alpha + \beta$.

The correspondence in question is the one defined by the tangent planes to B_0 at points of its triple curve on a given line in $\mathbb{P}^3$. This correspondence is clearly $(2\rho_t, 2\rho_t)$ - as Roth remarks - and the united points are due to the pinchpoints ν_2^t and to the points where the tangent line to the triple curve meets the line. There are ρ_t such tangent lines. Roth deduces the above formula from this, while it seems that, in this co respondence, since we're considering triples of points on the line, coincidence of all the three points should count 6 times and thus t correct formula should indeed be (IV_t) (in addition to the fact th its proof given for a j-tuple curve in general, seems to be OK). Moreover, Thm. 3 (IV) is in agreement with Noether's formulas.

5. Surfaces with a double and a cuspidal curve

Let Z be a smooth surface and $f: Z \to \mathbb{P}^3$ a finite map. Now we shall not assume that the ramification locus of f is finite. Thus $X = f(Z)$ might have a "cuspidal curve" - equal to the image b f of the 1-dimensional part of the ramification locus - in additio to isolated pinchpoints. Such a situation occurs for example if f is a nongeneric projection of an embedding $Z \subseteq \mathbb{P}^N$, or if f is th dual map of a smooth surface $Z \subseteq \check{\mathbb{P}}^3$.

Here we shall consider the case that X has only an "ordinary double curve and an "ordinary" cuspidal curve. To be precise, we shall assume $f: Z \to f(Z) = X \subseteq \mathbb{P}^3$ satisfies the conditions (1), (3 and (4) of Section 3, together with

(0)' char $k \neq 2, 3$

$(2)_{b+c}$: The singular locus of X is as follows.

There is a curve D_b of ordinary double points. At all but a finite number of points of D_b the condition (2) of Sectio 3 is satisfied. The remaining points of D_b lie also on the cuspidal curve D_c, D_c being the image of the 1-dimensional part of the ramification locus of f. At a general point $x \in D_c$, the completion of the local ring of X is isomorphi to

$$k[[t_1, t_2, t_3]] / (t_1^3 - t_2^2) .$$

The ramification ideal $F^0(\Omega^1_{Z/X})$ has no embedded components

Let $\varphi:\bar{Z} \to Z$ denote the blowup of the isolated ramification points of f, and set $E \subseteq \bar{Z}$ equal to the exceptional divisor. If $R \subseteq Z$ denotes the ramification scheme of f, defined by $F^0(\Omega^1_{Z/X})$, we let Γ_c be the 1-dimensional part of R (hence $\varphi^{-1}R = \varphi^{-1}\Gamma_c \cup E$). Let $\underline{C} = \underline{Hom}_{O_X}(f_* O_Z, O_X)$ denote the conductor of Z in X, and D (resp. Γ) the subscheme of X (resp. Z) defined by $\underline{C}$. From the above assumptions it follows that

$$[D] = [D_b] + [D_c]$$

holds in $A_1 X$. Now let Γ_b denote the reduced curve on Z satisfying

$$[\Gamma_b] = [f^{-1}\Gamma_b] \quad \text{in } A_1 Z$$

(beware that $f^{-1}\Gamma_b$ might have embedded components).

Lemma 4.

(i) $[\Gamma] = [\Gamma_b] + 2[\Gamma_c]$

(ii) $K_Z = (\mu_0 - 4)H - [\Gamma_b] - 2[\Gamma_c]$

(iii) $\check{h} \cap [\bar{Z}] = (\mu_0 - 1)H - [\varphi^{-1}\Gamma_b] - 3[\varphi^{-1}\Gamma_c] - [E]$.

Proof: Similar to the proofs of the lemmas of Section 3.

We shall now introduce the various numerical characters of the surface $X \subseteq \mathbb{P}^3$ (compare [S], p.314; [Z], p.450, but note that the situation here is simpler since we assume X has no isolated singularities).

The characters μ_0, μ_1, μ_2, κ, and $\nu_2 = -(E^2)$ are defined as before. We set

b = degree of D_b

q = rank of D_b

ρ = class of immersion of D_b in X

ν_b = # ramification points of $\Gamma_b \to D_b$

$\nu_b' = \#$ ramification points of $\Gamma_b' \to D_b'$

$\nu_b'' = \#$ ramification points of $D_b' \to D_b$

$\delta_b = g_a(\Gamma_b) - g(\Gamma_b)$.

c = degree of D_c

r = rank of D_c

σ = class of immersion of D_c in X

$\nu_c = \#$ ramification points of $\Gamma_c' \to D_c$

$\delta_c = g_a(\Gamma_c) - g(\Gamma_c)$

$i = (\Gamma_b, \Gamma_c)$.

The relations among the above characters that we are going to prove are stated in the next theorem. Unfortunately there is one relation we are unable to prove at the present time. We give it in a conjectural form (IV_c) together with another relation (III_c) we deduce from it, so as to get a picture of the complete set of relation

<u>Theorem 4</u>:

(I) $\mu_1 = \mu_0(\mu_0 - 1) - 2b - 3c$

(II) $\mu_1(\mu_0 - 2) = \kappa + \rho + 2\sigma$

(III_b) $b(\mu_0 - 2) = \rho + \delta_b + \nu_b'' + i$

(IV_b) $2\rho - 2q = \nu_b'$

(V) $\mu_2 + 2\nu_2 + \sigma = \mu_1 + \kappa$

<u>Conjecture</u>:

(III_c) $c(\mu_0 - 2) = 2\sigma + 2\delta_c + 2\nu_c + i$

(IV_c) $2\sigma - r = (\Gamma_c^2) - \nu_c$.

<u>Proof</u>. (I) follows from lemma 4 (iii). The proof of (V) goes as in thm. 2. Note that

$$\sigma = \int \check{h} \cap [\Gamma_c] \quad \text{(by definition), hence}$$

$$\sigma = (\mu_0 - 1)c - i - 3(\Gamma_c^2) \quad \text{(by lemma 4 (iii))}.$$

To prove (II) we compute $\mu_2 = \int \check{h}^2$, using again lemma 4 (iii). We obtain

$$\mu_2 = (\mu_0 - 1)\mu_1 - \rho - 3\sigma - 2\nu_2 ,$$

hich together with (V) yields (II).

The proof of (IV_b) is just like the proof of Thm. 2 (IV). From his and $(*_b)$ of the next lemma, (III_b) follows.

Lemma 5: $(*_b)$ $2q + 2\nu''_b + \nu'_b + 2\delta_b = 2b(\mu_0 - 2) - 2i$

$*_c)$ $r + \nu_c + 2\delta_c = c(\mu_0 - 2) - (\Gamma_c^2) - i$.

Proof: The adjunction formula for $\Gamma_b \subseteq Z$ gives (with lemma 4 ii))

$$\begin{aligned} 2g_a(\Gamma_b) - 2 &= \int (\mu_0 - 4)h \cap [\Gamma_b] - 2(\Gamma_b, \Gamma_c) \\ &= 2b(\mu_0 - 4) - 2i . \end{aligned}$$

y definition of δ_b , and by the Riemann-Hurwitz formula applied to he map $\Gamma'_b \to D'_b$ between the normalizations of Γ_b and D_b ,

$$2(2g(D_b) - 2) + \nu'_b + 2\delta_b = 2b(\mu_0 - 4) - 2i .$$

hen apply the formula for the rank, q , of D_b ,

$$q = 2b + 2g(D_b) - 2 - \nu''_b$$

o get $(*_b)$.

Similarly,

$$2g_a(\Gamma_c) - 2 = c(\mu_0 - 4) - i - (\Gamma_c^2) ,$$

$$r - 2c + \nu_c + 2\delta_c = c(\mu_0 - 4) - i - (\Gamma_c^2) .$$

Q.E.D.

Corollary 3: $(\Gamma_b^2) = \mu_0 b + \delta_b - \nu_2 - 2i$.

Proof: Compute $\rho + (\Gamma_b^2)$, using $\rho = \int \check{h} \cap ([\varphi^{-1}\Gamma_b] - [E])$ and emma 4 (iii). Then substitute for ρ,- using Thm. 4 (III_b).

Q.E.D.

Remark that Cor. 3 can be viewed as a formula for the singularities δ_b of Γ_b, and generalizes the triple point formula Prop. 2.

In the same manner we see that, given $(*_c)$ of Lemma 5, the conjectured formula (IV_c) is equivalent to

$(IV_c)'$ $6(\Gamma_c^2) = \mu_0 c + 2\delta_c + 2\nu_c - i$.

Again, $(IV_c)'$ can be viewed as formula for the singularities of Γ_c, and this suggests that there should exist a "triple point formula" for X, giving the "number of singularities" of the "doubl point scheme" of $f:Z \to X$ (recall that Γ was defined by the condu tor of Z in X), and that we should be able to obtain $(IV_c)'$ from such a formula.

Now Kleiman has a higher order multiple point theory (see [K] Ch. V, D), which he has generalized to local complete intersection ma (unpublished). It is easy to show (by reduction to hyperplane secti of X) that the formula giving the number of triple points of the m $f:Z \to X$ does apply in the present case. However, how to compute t triple points of f in terms of the characters given here, is not clear (this requires an analysis of the double point scheme of f, as to understand how much it differs from Γ; the problem is mainl how to "count" the intersections of Γ_b and Γ_c).

The formula would look like:

(T) $\quad \#$ triple points of $f = (\Gamma_b^2) + 4(\Gamma_c^2) + 4i - \mu_0(b+c) + c_2(\nu)$

where $c_2(\nu)$ denotes the degree of the 2nd Chern class of the virtu conormal bundle $f^*\Omega_{\mathbb{P}^3} - \Omega_Z^1$. A purely formal computation shows:

$$c_2(\nu) = c_1^2 - c_2 + 4\int h \cap K_Z + 6\mu_0 .$$

Since we have an expression for K_Z, we can find one for c_1^2, but it is not so clear how to find an expression for c_2. We prefer n to make further comments on this matter, but hope to report on the question later (see Appl. 3 below, however).

We shall now apply a result of ([P1]), as we did in Section 3 to obtain an expression for the cuspidal curve of the dual surface of X.

Let $\check{\pi}: \bar{Z} \to \check{X} = \check{\pi}(\bar{Z})$ denote the dual map. We let $R \subseteq \bar{Z}$ denote the ramification scheme of $f \circ \pi : \bar{Z} \to X$, so that R is defined by the ideal $F^0(\Omega^1_{\bar{Z}/X})$, and $R^* \subseteq \bar{Z}$ denotes the ramification scheme of $\check{\pi}$, defined by $F^0(\Omega^1_{\bar{Z}/\check{X}})$. Then ([P1], 3.4) yields a formula

$$[R^*] - [R] = 4(\check{h}-h) \cap [\bar{Z}] \quad \text{in} \quad A_1\bar{Z} .$$

<u>Proposition 6</u>: (i) $\deg(f\circ\pi)_*[R^*] = 4(\mu_0-2)\mu_0 - 8b - 11c$

(ii) $$\deg \check{\pi}_*[R^*] = 4\mu_0(\mu_0-1)(\mu_0-2) - 8b(2\mu_0-3) - c(23\mu_0-35) + 4(\Gamma_b^2) + 33(\Gamma_c^2) + 23i + 2\nu_2 .$$

<u>Proof</u>: (i) $$(f\circ\pi)_*[R^*] = 4(\check{h}-h) \cap [X] + (f\circ\pi)_*[R] = 4(\mu_0-2)H - 4[\Gamma_b] - 12[\Gamma_c] + [\Gamma_c] .$$

(ii) From the proof of ([P1], 3.4) it follows:

$$[R] = (c_1(\Omega^1_{\bar{Z}}) - \pi^* c_1(\Omega^1_Z)) \cap [\bar{Z}] + [\Gamma_c] + [E] = [\Gamma_c] + 2[E] .$$

Hence we get (use also lemma 4 (iii)):

$$[R^*] = 4(\mu_0-2)h \cap [\bar{Z}] - 4[\Gamma_b] - 11[\Gamma_c] + 2[E]$$

$$\deg \check{\pi}_*[R^*] = \int \check{h} \cap [R^*] = 4(\mu_0-2)(\mu_0-1)\mu_0 - 4(\mu_0-1)\cdot 2b - 11(\mu_0-1)c - 4(\mu_0-2)\cdot 2b + 4(\Gamma_b^2) + 11i - 12(\mu_0-2)c + 12i + 33(\Gamma_c^2) + 2\nu_2 ,$$

$$\deg \check{\pi}_*[R^*] = 4\mu_0(\mu_0-1)(\mu_0-2) - 8b(2\mu_0-3) - (23\mu_0-35)c + 4(\Gamma_b^2) + 33(\Gamma_c^2) + 23i + 2\nu_2 .$$

Q.E.D.

<u>Application 2</u> (see [R], p. 549): <u>Surface of contact</u>.

Suppose $Y \subseteq \mathbb{P}^N$ is a smooth threefold. In the same way that we defined the curve of contact $C \subseteq Z$ with respect to a generic projection $Z \to \mathbb{P}^2$ of a smooth surface Z, we define the surface of contact $A \subseteq Y$ to be the subscheme defined by the ramification ideal $F^0(\Omega^1_{Y/\mathbb{P}^3})$ of a generic projection $p: Y \to \mathbb{P}^3$. Set $A_0 = p(A)$. One can show that

A_0 has a double and a cuspidal curve in the sense of this section. Let us use Roth's notation: A_0 has degree a, rank m, a double curve of degree δ and containing ${}_3\mu_2$ triple points, a cuspidal curve of degree κ, meeting the double curve in μ_4 points which a stationary on the cuspidal curve and in μ_{3-2} points which are stationary on the double curve. In order to find an expression for the class, w, of A_0, he quotes a formula of Cayley (see [S], p. 319):

$$w = 3\,{}_3\mu_2 - 3\binom{a}{3} + \frac{3}{2}am - 2m + 8\mu_4 + \frac{9}{2}\mu_{3-2} .$$

If we use Thm. 4(I), this formula is the same as

$$w = a(a-1)^2 - \tfrac{1}{2}(2\delta+3\kappa)(3a-4) + 3\,{}_3\mu_2 + 8\mu_4 + \frac{9}{2}\mu_{3-2} .$$

Assume the following:

(a) the conjecture (III_c) holds

(b) the intersection number i of the pullbacks to A of the double and the cuspidal curve of A_0 is equal to $2\mu_4 + \mu_{3-2}$.

Then the above formula for the class w is a corollary of Thm. 4.

Application 3: The dual of a smooth surface

The preceding results apply in particular to the case that $X \subseteq \mathbb{P}^3$ is the dual surface of a smooth surface in $\check{\mathbb{P}}^3$. So assume $Z \subseteq \check{\mathbb{P}}^3$ is a smooth (non-linear) surface and let $f: Z \to f(Z) = X \subseteq \mathbb{P}^3$ denote its dual map (which we assume to be separable).

Lemma 6: The dual map $f: Z \to \mathbb{P}^3$ of the smooth surface $Z \subseteq \check{\mathbb{P}}^3$ is finite.

Proof: Suppose $E \subseteq Z$ is a curve. Then $\int \check{h} \cap [E] > 0$. Since $h = (n-1)\check{h}$, where n = degree of Z in $\check{\mathbb{P}}$, it follows that $\int h \cap [E] > 0$, i.e., $\deg f_*[E] > 0$, hence f does not collapse E. Q.E.D.

We shall now compute the characters of X in terms of the degree n of Z in $\check{\mathbb{P}}^3$; we keep the notation of this section.

<u>Lemma 7</u>. (i) $[\Gamma_c] = 4(n-2)\check{H}$

(ii) $[\Gamma_b] = (n-2)(n^3-n^2+n-12)\check{H}$.

<u>Proof</u>: (i) follows from the proof of Prop. 6.
To prove (ii), we use Lemma 4 (iii) together with (i), and the fact that $\mu_0 = n(n-1)^2$ holds (Thm. 1, (i)).

Q.E.D.

In order to show that we actually obtain Salmon-Cayley-Zeuthen's formulas for the characters of $X \subseteq \mathbb{P}^3$, we have to make two assumptions - one is that Conj. (IV_c) holds, and the other has to do with how the singularities of $\Gamma \subseteq Z$ are. Namely, we assume $\delta_c = 0$ and $\nu_c = \nu'_b = \beta$ say, and that if we set $\gamma = \nu''_b$, then we have $\delta_b = 3t + \gamma$ and $i = 2\beta + \gamma$. Note that these last assumptions are in agreement with the geometric description of the singularities of X given by Salmon-Cayley-Zeuthen ([S],[Z]). We shall not, however, try to justify them further here - remark only that they follow if one shows that the intersections of D_b and D_c are of two types: "type γ " (see Fig. 1) and "type β " (see Fig. 2).

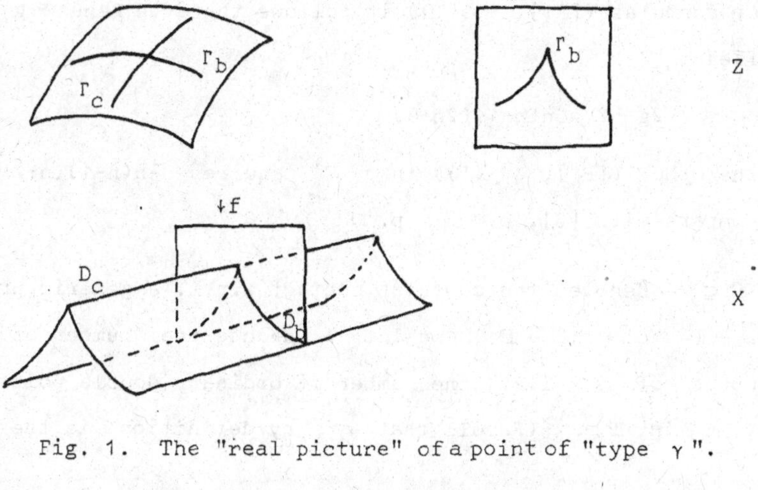

Fig. 1. The "real picture" of a point of "type γ ".

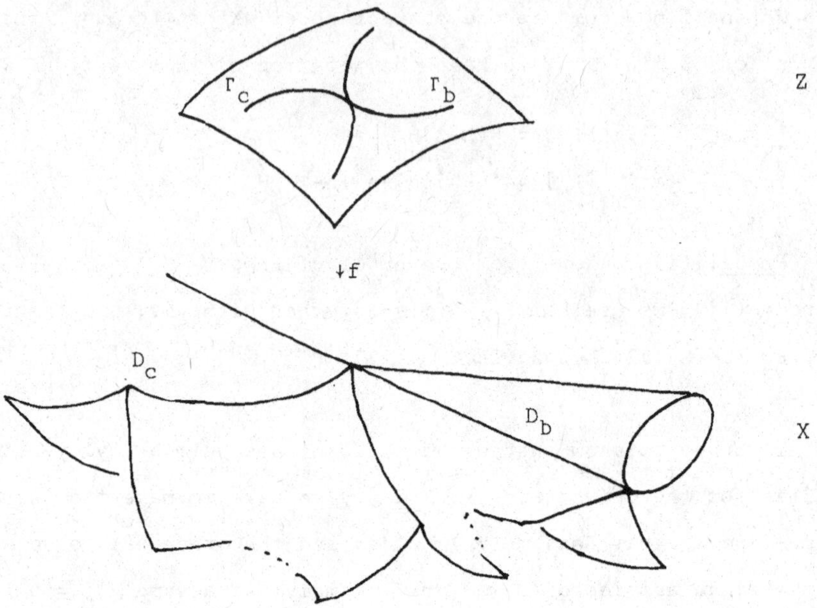

Fig. 2. The "real picture" of a point of "type β".

Recall that n denotes the degree of Z in $\check{\mathbb{P}}^3$. From Thm. it follows, since Z is smooth, that Z has rank n(n-1), its class is $n(n-1)^2$, and if $C^* \subseteq Z$ is the curve of contact w.r.t. generic projection $Z \to \check{\mathbb{P}}^2$ then C (which has degree n(n-1) and class $n(n-1)^2$) aquires n(n-1)(n-2) cusps in $\check{\mathbb{P}}^2$. From one of the Plücker formulas ([P1], 3.9(II)) it follows that the genus $\check{g}$ c C^* satisfies

$$2g - 2 = n(n-1)(2n-5) ,$$

and from the other ([P1], 3.9(I)) that C^* aquires $\frac{1}{2}n(n-1)(n-2)(n$- nodes. (Compare with [B], p. 150, p. 155.)

If $C \subseteq Z$ denotes the curve of contact w.r.t. a generic projection $Z \to \mathbb{P}^2$ of $f: Z \to \mathbb{P}^3$, we let δ denote the "number of apparent nodes" of C, i.e., the number of ordinary double points aquired by C in $\mathbb{P}^2$. (Recall that κ, by definition, is the number of cusps aquired by C .) Likewise, we let k (resp. h) denote the number of apparent nodes of Γ_b (resp. Γ_c). From our assumptions it follows that Γ_c is smooth ($\delta_c = 0$), that $\Gamma_c \to D_c$ has β ramification points (whose images are cusps on D_c, see Fig. 2), that D_b and Γ_b have γ cusps, while D_b has t triple points and Γ_b has $3t$ double points, $\Gamma_b \to D_b$ has $\gamma + \beta$ ramifiction points, while $\Gamma_b' \to D_b'$ ramifies at β points and $D_b' \to D_b$ at γ points.

Applying Thm. 4, and the conjecture, as well as the Plücker formulas (as referred to above for C^*) to find δ, k, and h, we obtain the following expressions, which agree with Salmon-Cayley-Zeuthen's ([S], p. 602).

(1) $\mu_0 = n(n-1)^2$

(2) $\mu_1 = n(n-1)$

(3) $\mu_2 = n$

(4) $\delta = \frac{1}{2}n(n-2)(n^2-9)$

(5) $\kappa = 3n(n-2)$

(6) $b = \frac{1}{2}n(n-1)(n-2)(n^3-n^2+n-12)$

(7) $k = \frac{1}{8}n(n-2)(n^{10}-6n^9+16n^8-54n^7+164n^6-288n^5+547n^4-1058n^3+1068n^2$
$-1214n+1464)$

(8) $t = \frac{1}{6}n(n-2)(n^7-4n^6+7n^5-45n^4+114n^3-111n^2+548n-960)$

(9) $q = n(n-2)(n-3)(n^2+2n-4)$

(10) $\rho = n(n-2)(n^3-n^2+n-12)$

(11) $c = 4n(n-1)(n-2)$

(12) $h = n(n-2)(8n^4-32n^3+40n^2-54n+78)$

(13) $r = 2n(n-2)(3n-4)$

(14) $\sigma = 4n(n-2)$

(15) $\beta = 2n(n-2)(11n-24)$

(16) $\gamma = 4n(n-2)(n-3)(n^3+3n-16)$

Remark 5: The formulas (6), (8), and (16) are also proved in ([V], 8.5(ii),(iii), 9.5.1(ii)) - the approach there being the stud of order of contact of Z with lines and planes in $\check{\mathbb{P}}^3$.

As remarked earlier, the triple point formula (T) applies in the present case. Moreover, one is able to compute the right hand side of (T):

$$(T)_n \quad \#\text{triple points of } X = \tfrac{1}{2}n(n^8-6n^7+15n^6-35n^5+84n^4-123n^3+158n^2 -256n+192)$$

In fact, we use Lemma 7 for the first two terms, then (1), (6 and (11) for the next, and finally Prop. 1 (i), (ii), and Lemma 4 (i to compute

$$c_2(\nu) = n(n-4)^2 - n(n^2-4n+6) + 4n(n-1)(n-4) + 6n(n-1)^2$$
$$= 2n(5n^2-18n+16)$$

Since the obtained expression for the number of triple points also turns out to be equal to

$$3t + 3\beta + 3\gamma ,$$

using (8), (15), and (16), it seems reasonably clear that this is in fact the right way to count the triple points of X, though again we do not try to justify further the coefficients of β and γ her

Example 3. A smooth cubic.

Suppose $n = 3$. The double curve D_b of X has degree $b = 27$, and its class of immersion (by definition equal to the degree of Γ_b in $\check{\mathbb{P}}^3$) is also 27 - in fact $\Gamma_b \subseteq Z$ is equal to the union of the 27 lines on Z. And D_b has 45 triple points while D_c has 54 cusps, so that there are 99 triple points altogether on (there are no points of "type γ").

The curve C^* has genus 4, while C has genus 1. The

cuspidal curve D_c has genus 19.

(See [S], pp. 184. Salmon was led to his theory of reciprocal surfaces by a study of the reciprocal of a cubic. He says "The first attempt to explain the effect of nodal and cuspidal lines on the degree of the reciprocal was made in the year 1847, in two papers which I contributed to the Cambridge and Dublin Mathematical Journal, II, p. 65, and IV, p. 188. It was not till the close of the year 1849, however, that the discovery of the twenty-seven right lines on a cubic by enabling me to form a clear conception of the nature of the reciprocal of a cubic, led me to the theory in the form here explained ..." ([S], Footnote on p. 301).)

Example 4. A smooth quartic.

Suppose $n = 4$ (as is well known, Z is then a K3 surface).

The double curve D_b has degree $b = 480$, $\gamma = 1920$ cusps, and $t = 3200$ triple points. It has genus 561.

The cuspidal curve has degree $c = 96$, $\beta = 320$ cusps, and genus 129.

In all, X has $t + \beta + \gamma = 5440$ triple points!

The curve of contact $C^* \subseteq Z$ with respect to a generic projection $Z \to \check{\mathbb{P}}^2$ has genus 19, while the curve of contact $C \subseteq Z$ w.r.t. a generic projection $Z \to \mathbb{P}^2$ has genus 9.

Bibliography

[A] M. Artin, On Enriques Surfaces, Doctoral thesis, Harvard University, 1960.

[B] H.F. Baker, Principles of geometry. Vol.VI. Introduction to the theory of algebraic surfaces and higher loci. Cambridge Univ.Press, 1933.

[F] W. Fulton, "Rational equivalence on singular varieties". Publ.Math.IHES, 45 (1976), Paris.

[K-L] G. Kempf, D.Laksov, "The determinantal formula of Schubert calculus". Acta Math., 132 (1974), 153-162.

[K] S. Kleiman, "The enumerative theory of singularities". Proceedings of the Nordic Summer School, Oslo 1976(Noordhoff)

[La] D. Laksov, "Secant bundles and Todd's formula for the double points of maps into $\mathbb{P}^n$". Preprint, M.I.T., 1976.

[Ll] E. Lluis, "De las singularidades que aparacen al proyectar variedades algebraicas". Bol.Soc.Mat. Mexicana, Ser. 2, 1 (1956), 1-9.

[N] M. Noether, "Sulle curve multiple di superficie algebriche". Ann.d.Mat., 5 (1871-3), 163-177.

[P1] R. Piene, "Polar classes of singular varieties". Ann. scient. Éc. Norm. Sup. t. 11, 1978, fasc. 2.

[P2] R. Piene, "A proof of Noether's formula for the arithmetic genus of an algebraic surface". Preprint, M.I.T., 1977.

[P3] R.Piene, "Numerical characters of a curve in projective n-space". Proceedings of the Nordic Summer School, Oslo 19 (Noordhoff).

[R1] J. Roberts, "Generic projections of algebraic varieties". Am.J.Math., 93 (1971), 191-215.

[R2] J. Roberts, "Singularity subschemes and generic projections". Trans.AMS, 212 (1975), 229-268.

[R3] J. Roberts, "Variations of singular cycles in an algebraic family of morphisms". Trans.AMS, 168 (1972), 153-164.

[R] L. Roth, "Some formulae for primals in four dimensions". Proc. London Math.Soc., Ser. 2, 35 (1933), 540-550.

[S] G. Salmon, A treatise on the analytic geometry of three dimensions. Vol. II, 5th ed., Dublin 1915.

[SAG 7] N. Katz, "Pinceaux de Lefschetz: théorème d'existence". Exp. XVII in SGA 7, II, Springer L.N.M., 340.

[S-R] J.G. Semple, S. Roth, Introduction to algebraic geometry. Oxford 1949.

[V] I. Vainsencher, On the formula of de Jonquières for multiple contacts. Doctoral thesis, M.I.T., 1976.

[W] A. Wallace, "Tangency and duality over arbitrary fields". Proc. London Math.Soc., Ser. 3, 6 (1956), 321-342.

[Z] H.G. Zeuthen, "Révision et extension des formules numériques de la théorie des surfaces réciproques". Math.Ann., 10 (1876), 446-546.

Institute of Mathematics
University of Oslo
P.O.Box 1053 Blindern
Oslo 3, Norway

UN THEOREME DE STRUCTURE LOCALE POUR LES COMPLEXES PARFAITS

Marguerite FLEXOR - Lucien SZPIRO

0. Introduction.

On connaît le théorème suivant de A. Grothendieck (cf. [2]) : soient $X \xrightarrow{f} \mathrm{Spec}\, A$ un morphisme propre où A est un anneau (commutatif) noethérien, $\mathcal{F}$ un faisceau cohérent sur X plat sur A , il existe un complexe borné $\ldots 0 \longrightarrow L^0 \longrightarrow L^1 \longrightarrow \ldots \longrightarrow L^n \longrightarrow 0 \ldots$ de A-modules projectifs de rang fini tel que l'on ait fonctoriellement, pour tout A-module N et pour tout $i \in \mathbb{Z}$, un isomorphisme : $R^i f_*(\mathcal{F} \otimes f^*N) \simeq H^i(L^\cdot \otimes N)$.

Nous démontrons ici une réciproque de ce théorème, réciproque qui d'ailleurs avait été conjecturée (*) dans [2] , (théorème 1). La fonctorialité de la construction que nous faisons, permet d'obtenir le théorème 1' qui peut s'interpréter comm une puissance tensorielle du théorème 1 .

Notons que le cas où $\mathcal{L}^\cdot$ est le complexe $0 \longrightarrow A \xrightarrow{f} A \longrightarrow 0$ sur un anneau A a déjà été traité par D. Mumford ([3] page 53), comme nous l'a signalé D. Ferrand. Il se trouve que la construction de D. Mumford (loc. cit.) est en quelque sorte duale de la nôtre.

Le théorème 1" réinterprète le théorème 1 en terme du cône relatif sur $\mathbb{P}^n_A$. Ceci nous permet d'obtenir, si M est un module de dimension projective finie égale à n et N un A-module de type fini, que $\mathrm{Tor}_i^A(M,N)$ est la partie homogè de degré (n + 1) de l'homologie $H_i(\underline{X},\cdot)$ d'un complexe de Koszul gradué. Nous espérons que les spécialistes (?) de la conjecture des Tor apprécieront.

Notons que le lecteur algébriste, ou qui ne se préoccupe pas de morphismes projectifs peut directement lire le pagraphe II .

(*) par A. Grothendieck.

Pour fixer les notations, si $M = \bigoplus_{n \in \mathbb{Z}} M_n$ est un module gradué sur un anneau gradué A, nous noterons, si $k \in \mathbb{Z}$, $M(k)$ le A-module gradué dont la partie homogène de degré n est donnée par $M(k)_n = M_{n+k}$.

D'autre part, si $\mathcal{L}^\bullet$ est un complexe de modules sur un schéma S et k un entier, nous noterons $\mathcal{L}^\bullet[k]$ le translaté k fois à gauche.

I. Un inverse à droite du foncteur $R p_*$ sur l'espace projectif.

Soit $\mathcal{L}^\bullet : 0 \longrightarrow L^0 \longrightarrow L^1 \longrightarrow \dots \longrightarrow L^n \longrightarrow \dots$ un complexe de faisceaux localement libres de rang fini sur un schéma noethérien S. Nous dirons que $\mathcal{L}^\bullet$ est de longueur $\leqslant n$ si $L^j = 0$ pour $j > n$. Nous désignerons par $\underline{C}_n$ la catégorie des complexes de faisceaux localement libres de rang fini de longueur $\leqslant n$. Notons $P = \mathbb{P}^n_S$ l'espace projectif relatif sur S, f sa projection canonique, dont les fibres sont de dimension n, et $K_\bullet$ le complexe de Koszul sur P construit à partir de l'homomorphisme $\mathcal{O}_P(-1)^{n+1} \longrightarrow \mathcal{O}_P$. Par définition on a

$$K_i = f^*(\bigwedge^i \mathcal{O}_S^{n+1}) \otimes_P \mathcal{O}_P(-i) .$$

Pour tout objet $\mathcal{L}^\bullet$ de $\underline{C}_n$, on peut construire le complexe produit tensoriel $\mathcal{C}(\mathcal{L}^\bullet) = K_\bullet \otimes f^*\mathcal{L}_\bullet[n]$ et le sous-complexe $\mathcal{C}'(\mathcal{L}^\bullet)$ de $\mathcal{C}(\mathcal{L}^\bullet)$:

$$\mathcal{C}'(\mathcal{L}^\bullet) : 0 \longrightarrow \mathcal{C}_n(\mathcal{L}^\bullet) \longrightarrow \mathcal{C}_{n-1}(\mathcal{L}^\bullet) \longrightarrow \dots \longrightarrow \mathcal{C}_0(\mathcal{L}^\bullet) \longrightarrow 0 .$$

Soit $\mathcal{F}(\mathcal{L}^\bullet) = \mathrm{Ker}(\mathcal{C}_n(\mathcal{L}^\bullet) \longrightarrow \mathcal{C}_{n-1}(\mathcal{L}^\bullet))$. A un morphisme $u : \mathcal{L}^\bullet \longrightarrow \mathcal{M}^\bullet$ de $\underline{C}_n(S)$, correspond de manière naturelle des morphismes :

$$\mathcal{C}'(u) : \mathcal{C}'(\mathcal{L}^\bullet) \longrightarrow \mathcal{C}'(\mathcal{M}^\bullet)$$

et

$$\mathcal{F}(u) : \mathcal{F}(\mathcal{L}^\bullet) \longrightarrow \mathcal{F}(\mathcal{M}^\bullet) .$$

LEMME 1.- Avec les notations introduites ci-dessus on a :

a) Le faisceau $\mathcal{F}(\mathcal{L}^\bullet)$ est localement libre.

b) Si $u : \mathcal{L}^\bullet \longrightarrow \mathcal{M}^\bullet$ et $v : \mathcal{M}^\bullet \longrightarrow \mathcal{N}^\bullet$ sont des morphismes de $\underline{C}_n(S)$, alors $\mathcal{F}(u \circ v) = \mathcal{F}(u) \circ \mathcal{F}(v)$.

Autrement dit, $\mathcal{L}^\bullet \longrightarrow \mathcal{F}(\mathcal{L}^\bullet)$ est un foncteur de $\underline{C}_n(S)$ dans $\underline{C}_0(\mathbb{P}^n_S)$.

Montrons a) : Comme $\mathcal{H}_i(K.) = 0$, pour tout i , $\mathcal{H}_i(\mathcal{C}(\mathcal{L}^\bullet)) = 0$ pour tout i . En particulier la suite

$$0 \longrightarrow \mathcal{F}(\mathcal{L}^\bullet) \longrightarrow \mathcal{C}_n(\mathcal{L}^\bullet) \longrightarrow \dots \longrightarrow \mathcal{C}_0(\mathcal{L}^\bullet) \longrightarrow 0$$

est exacte et $\mathcal{F}(\mathcal{L}^\bullet)$ est localement libre. Il est clair que l'on a c) .

Notons ρ le morphisme canonique de $\underline{C}_n(S)$ dans la catégorie dérivée $\underline{D}^b($ des complexes bornés de faisceaux localement libres sur S .

THEOREME.- Le diagramme suivant est commutatif :

$$\begin{array}{ccccc} \text{(complexes parfaits)} & \underline{C}_n(S) & \xrightarrow{\ F\ } & C_0(\mathbb{P}^n_S) & \text{(faisceaux localement libres)} \\ & & \searrow^{\rho} \quad \swarrow^{Rp_*} & & \\ & \text{(catégorie dérivée)} & \underline{D}^b(S) & & \end{array}$$

Soit $\mathcal{L}^\bullet$ un objet de $\underline{C}_n(S)$. Le théorème résulte des quatre points suiva

1) On a $\mathcal{F}(\mathcal{L}^\bullet)[n] = \mathcal{C}'(\mathcal{L}^\bullet)$ dans la catégorie dérivée de $\mathbb{P}^n_S$.

2) $R^i f_*(\mathcal{C}_j(\mathcal{L}^\bullet)) = 0$ pour $i > 0$, $0 \leqslant j \leqslant n$. En effet

$$\mathcal{C}_j = \bigoplus_{L=0}^{j} [f^*(\overset{i}{\wedge} \mathcal{O}_S^{n+1}) \otimes \mathcal{O}_P(-i) \otimes f^* L_{i-j}[n]] \ .$$

3) $f_*(\mathcal{C}_j) = L^{n-j}$ pour tout $0 \leqslant j \leqslant n$.

4) Si $u : \mathcal{L}^\bullet \longrightarrow \mathcal{M}^\bullet$, $u = Rp_*(\mathcal{C}'(u)) = Rp_*(\mathcal{F}(u))$ dans la catégorie dérivée $\underline{D}^b(S)$.

PROPOSITION.- Le foncteur $\mathcal{F}$ est exact, i.e. si $0 \longrightarrow \mathcal{N}^\bullet \xrightarrow{u} \mathcal{L}^\bullet \xrightarrow{v} \mathcal{M}^\bullet \longrightarrow$ est une suite exacte d'objets de $\underline{C}_n(S)$; on a la suite exacte suivante de fais

$$0 \longrightarrow \mathcal{F}(\mathcal{N}^\bullet) \xrightarrow{\mathcal{F}(u)} \mathcal{F}(\mathcal{L}^\bullet) \xrightarrow{\mathcal{F}(v)} \mathcal{F}(\mathcal{M}^\bullet) \longrightarrow 0 \ .$$

En effet, on a une suite exacte de complexes sur $\mathbb{P}^n_S$:

$$0 \longrightarrow \mathcal{C}'(\mathcal{N}^\bullet) \xrightarrow{\mathcal{C}'(u)} \mathcal{C}'(\mathcal{L}^\bullet) \xrightarrow{\mathcal{C}'(v)} \mathcal{C}'(\mathcal{M}^\bullet) \longrightarrow 0$$

et une suite exacte d'homologie :

$$0 \longrightarrow \mathcal{H}_n(\mathcal{C}^\cdot(\mathcal{N}^\cdot)) \longrightarrow \mathcal{H}_n(\mathcal{C}^\cdot(\mathcal{L}^\cdot)) \longrightarrow \mathcal{H}_n(\mathcal{C}^\cdot(\mathcal{M}^\cdot)) \longrightarrow \mathcal{H}_{n-1}(\mathcal{C}^\cdot(\mathcal{N}^\cdot))$$

soit encore :

$$0 \longrightarrow \mathcal{F}(\mathcal{N}^\cdot) \longrightarrow \mathcal{F}(\mathcal{L}^\cdot) \longrightarrow \mathcal{F}(\mathcal{M}^\cdot) \longrightarrow \mathcal{H}_{n-1}(\mathcal{C}^\cdot(\mathcal{N}^\cdot))$$

et la démonstration de a) dans le lemme 1 montre que $\mathcal{H}_{n-1}(\mathcal{C}^\cdot(\mathcal{N}^\cdot)) = 0$.

D'où la proposition.

COROLLAIRE.- Soient S un schéma noethérien, L et M des faisceaux localement libres de rang fini sur S , $P = \mathbb{P}^1_S \xrightarrow{p} S$ la droite projective sur S et ω le faisceau dualisant relatif à p $(\omega \simeq \mathcal{O}_P(-2))$. Il existe un isomorphisme naturel

$$\pi : \mathcal{H}om_S(L,M) \longrightarrow \mathcal{E}xt^1_P(p^*L, p^*M \otimes \omega)$$

défini de la manière suivante :

si $d \in \mathcal{H}om_S(L,M)$, $\pi(d) = \mathcal{F}(0 \longrightarrow L \longrightarrow M \longrightarrow 0)$.

1) Construction de π .- Notons $\mathcal{M}^\cdot : 0 \longrightarrow 0 \longrightarrow M \longrightarrow 0$, $\mathcal{N}^\cdot : 0 \longrightarrow L \longrightarrow 0 \longrightarrow 0$ les complexes de longueur $\leqslant 1$. On a une suite exacte de complexes de longueur $\leqslant 1$,

$$0 \longrightarrow \mathcal{M}^\cdot \longrightarrow \mathcal{L}^\cdot \longrightarrow \mathcal{N}^\cdot \longrightarrow 0$$

et grâce à la proposition 1 , une suite exacte de faisceaux localement libres sur P :

$$0 \longrightarrow \mathcal{F}(\mathcal{M}^\cdot) \longrightarrow \mathcal{F}(\mathcal{L}^\cdot) \longrightarrow \mathcal{F}(\mathcal{N}^\cdot) \longrightarrow 0 .$$

Par construction, $\mathcal{F}(\mathcal{M}^\cdot) = p^*M \otimes \omega$ et $\mathcal{F}(\mathcal{N}^\cdot) = p^*L$, i.e. $\mathcal{F}(\mathcal{L}^\cdot) \in \mathcal{E}xt^1_P(p^*L, p^*M \otimes \omega)$

2) Considérons la suite exacte attachée à un élément $\mathcal{E}$ de $\mathcal{E}xt^1_P(p^*L, p^*M \otimes \omega)$:

$$0 \longrightarrow p^*M \otimes \omega \longrightarrow \mathcal{E} \longrightarrow p^*L \longrightarrow 0 .$$

Soit $d \in \mathcal{H}om_S(L,M)$ l'homomorphisme composé :

$$d : L = p_*p^*L \longrightarrow R^1p_*(p^*M \otimes \omega) = M \otimes R^1p_*(\omega) \longrightarrow M$$

et

$$\tilde{\pi} : \mathcal{E}xt^1(p^*L, p^*M \otimes \omega) \longrightarrow \mathcal{H}om_S(L,M)$$

défini par $\tilde{\pi}(\mathcal{E}) = d$.

3) On vérifie aisément que π n'est autre que le composé des isomorphismes suivants tous canoniques :

$$\mathcal{H}om_S(L,M) \longrightarrow \mathcal{H}om_S(L, M \otimes R^1p_*(\omega))$$

$$\mathcal{H}om_S(L, M \otimes R^1p_*(\omega)) \longrightarrow R^1p_*(\mathcal{H}om_P(p^*L, p^*M \otimes \omega)$$

$$R^1p_*(\mathcal{H}om_P(p^*L, p^*M \otimes \omega)) \longrightarrow \mathcal{E}xt^1_P(p^*L, p^*M \otimes \omega)$$

et que, d'autre part, $\tilde{\pi} = \pi^{-1}$.

Soient k un entier $\geqslant 1$ et $\mathcal{L}^\bullet$ un k-uple complexe de taille $[0,n_1]\times \ldots \times$ i.e. un $\mathbb{Z}^k$-objet gradué $(L_{i_1,\ldots,i_k})$ en faisceaux localement libres de rang fi sur S tel que $L_{i_1,\ldots,i_k} = 0$ s'il existe $j \in [1,\ldots,k]$ tel que $i_j > n_j$, muni pour $1 \leqslant i \leqslant k$ d'une différentielle d_i de multi degré $(0,\ldots,1,\ldots,0)$ (1 à la i-ème place) telle que $d_i \circ d_i = 0$ et $d_i d_j = d_j \circ d_i$ si $i \neq j$. Nous signerons par $\underline{C}_{n_1,\ldots,n_k}(S)$ la catégorie des k-uples complexes sur S de tail $[0,n_1] \times \ldots \times [0,n_k]$. Nous noterons le foncteur "complexe simple associé" :

$$\int : \underline{C}_{n_1,\ldots,n_k}(S) \longrightarrow \underline{C}_n(S)$$

où l'on a posé

$$n = \sum_{i=1}^{k} n_i \quad .$$

THEOREME 1'.- <u>Il existe un foncteur</u> $\mathcal{F}' : \underline{C}_{n_1,\ldots,n_k}(S) \longrightarrow \underline{C}_0(\mathbb{P}_S^{n_1} \times \ldots \times \mathbb{P}_S^{n_k})$ <u>tel que le diagramme suivant soit commutatif</u> :

$$\begin{array}{ccc} \underline{C}_{n_1,\ldots,n_k}(S) & \xrightarrow{\mathcal{F}'} & \underline{C}_0(\mathbb{P}_S^{n_1} \times \ldots \times \mathbb{P}_S^{n_k}) \\ & {\scriptstyle \rho\circ\int} \searrow \quad \swarrow {\scriptstyle Rp_*} & \\ & \underline{D}^b(S) & \end{array}$$

<u>où</u> $p : \mathbb{P}_S^{n_1} \times \ldots \times \mathbb{P}_S^{n_k} \longrightarrow S$ <u>est la projection canonique</u>.

Ceci se démontre évidemment par récurrence sur k . Si $k = 1$, c'est le
ıéorème 1 . Supposons donc $k > 1$.

Pour tout k-uple complexe $\mathcal{L}^\bullet$ de $[0,n_1] \times \dots \times [0,n_k]$, pour tout entier i ,
$\leqslant i \leqslant n_1$, désignons par $\mathcal{L}_i^\bullet$ le $k-1$ uple complexe de taille $[0,n_2] \times \dots \times [0,n_k]$

$$((\mathcal{L}_i^\bullet)_{j_2,\dots,j_k}) = (L_{i,j_2,\dots,j_k}) \qquad \begin{matrix} 0 \leqslant j_1 \leqslant n_1 \\ 2 \leqslant l \leqslant k \end{matrix} \quad .$$

ı obtient un complexe $\mathcal{D}^\bullet$ de $(k-1)$-uples complexes :

$$0 \longrightarrow \mathcal{L}_1^\bullet \longrightarrow \mathcal{L}_2^\bullet \longrightarrow \dots \longrightarrow \mathcal{L}_{n_1}^\bullet \longrightarrow 0$$

t le complexe $\int \mathcal{D}^\bullet$:

$$0 \longrightarrow \int \mathcal{L}_1^\bullet \longrightarrow \int \mathcal{L}_2^\bullet \longrightarrow \dots \longrightarrow \int \mathcal{L}_n^\bullet \longrightarrow 0$$

ınt le complexe simple associé n'est autre que $\int \mathcal{L}^\bullet$.

Par l'hypothèse de récurrence, il existe un foncteur :

$$\mathcal{F}' : \underline{C}_{n_2,\dots,n_k}(S) \longrightarrow \underline{C}_0(\mathbb{P}_S^{n_2} \times \dots \times \mathbb{P}_S^{n_k})$$

ɛl que, si $p : \mathbb{P}_S^{n_2} \times \dots \times \mathbb{P}_S^{n_k} \longrightarrow S$ est la projection canonique, le diagramme
ıivant soit commutatif :

$$\begin{array}{ccc} \underline{C}_{n_2,\dots,n_k}(S) & \longrightarrow & \underline{C}_0(\mathbb{P}_S^{n_2} \times \dots \times \mathbb{P}_S^{n_k}) \\ & {\scriptstyle p_\bullet \int} \searrow \quad \swarrow {\scriptstyle Rp_*} & \\ & D^b(S) & \end{array}$$

ı a donc sur $\mathbb{P}_S^{n_2} \times \dots \times \mathbb{P}_S^{n_k}$ un complexe $\mathcal{F}'(\mathcal{D}^\bullet)$ de faisceaux localement
.bres de rang fini :

$$0 \longrightarrow \mathcal{F}'(\mathcal{L}_1^\bullet) \longrightarrow \dots \longrightarrow \mathcal{F}'(\mathcal{L}_{n_1}^\bullet) \longrightarrow 0$$

ɛl que $Rp_*(\mathcal{F}'(\mathcal{D}^\bullet)) = \int \mathcal{D}^\bullet$ dans la catégorie dérivée $D^b(S)$.

ıit $\mathcal{F}'' : \underline{C}_n(\mathbb{P}_S^{n_2} \times \dots \times \mathbb{P}_S^{n_k}) \longrightarrow \underline{C}_0(\mathbb{P}_S^{n_1} \times \dots \times \mathbb{P}_S^{n_k})$ le foncteur défini au
ıéorème 1 .

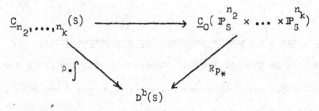

On vérifie aisément que le foncteur

$$\mathcal{F} : \mathcal{L}^{\bullet} \longrightarrow \mathcal{F}''(\mathcal{F}'(\mathcal{D}^{\bullet}))$$

satisfait au théorème 1' .

<u>Remarque</u>.- On peut considérer la catégorie $\underline{C}_n(S)$ comme une sous-catégorie de $\underline{C}_{n_1,\dots,n_k}(S)$, pour $\sum_{i=1}^{k} n_i = n$. En effet, il suffit d'écrire un complexe $\mathcal{L}^{\bullet} : 0 \longrightarrow L^0 \longrightarrow \dots \longrightarrow L^n \longrightarrow 0$ de $\underline{C}_n(S)$ de la manière suivante :

$$\begin{array}{ccccccccccccc}
0 & \longrightarrow & L^0 & \longrightarrow & \dots & \longrightarrow & L^{n_1-1} & \longrightarrow & L^{n_1} & \longrightarrow & 0 & & \\
 & & \downarrow & & & & \downarrow & & \downarrow d & & & & \\
0 & \longrightarrow & 0 & \longrightarrow & \dots & \longrightarrow & 0 & \longrightarrow & L^{n_2} & \longrightarrow & L^{n_2+1} & & \\
 & & & & & & & & & & \downarrow & \vdots & 0 \\
 & & & & & & & & & & \vdots & \downarrow & \downarrow \\
 & & & & & & & & & \longrightarrow & L^{n_{k-1}+1} & \longrightarrow \dots \longrightarrow & L^{n_k}
\end{array}$$

et nous pouvons donc appliquer le théorème 1' à $\underline{C}_n(S)$: il existe donc fonctor ment un faisceau localement libre $\mathcal{F}$ sur $\mathbb{P}_S^{n_1} \times \dots \times \mathbb{P}_S^{n_k}$ tel que $Rp_* \mathcal{F} = \mathcal{L}^{\bullet}$ la catégorie dérivée.

II. <u>Le cône relatif, cas d'un module de dimension projective finie</u>.

Soit M un module de type fini et de dimension projective finie sur un ann noethérien A . Soient $0 \longrightarrow L_n \longrightarrow \dots \longrightarrow L_0 \longrightarrow M \longrightarrow 0$, une ré lution projective finie de M et $\mathcal{L}.$ le complexe :

$$0 \longrightarrow L_n \xrightarrow{d_{n-1}} \dots \xrightarrow{d_0} L_0 \longrightarrow 0 \ .$$

Posons $S = A[X_0,\dots,X_n]$, $p : \operatorname{Spec} S \longrightarrow \operatorname{Spec} A$ la projection canonique et $s : \operatorname{Spec} A \longrightarrow \operatorname{Spec} S$ la section nulle. Soit $K.$ le complexe de Koszul de S sur A et $\mathcal{E}. = K. \otimes_S p^*\mathcal{L}. = 0 \longrightarrow E_{2n+1} \longrightarrow E_{2n} \longrightarrow \dots \dots \longrightarrow E_0 \longrightarrow 0$

On a :

$$H_i(\mathcal{E}.) = 0 \ \text{pour} \ i > 0 \ \text{et} \ H_0(\mathcal{E}.) = s^*M \ .$$

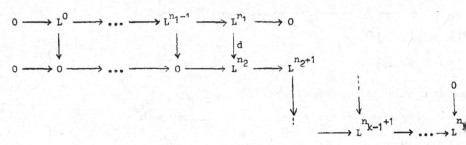

En effet :

$$H_i(\mathcal{E}.) = \mathrm{Tor}_i^S(S/\underline{X}, M[X_0 \dots X_n]) \ .$$

Posons $\mathcal{M}$ = Ker $E_n \longrightarrow E_{n-1}$, d'où une suite exacte de S-modules :

$$(*) \qquad 0 \longrightarrow \mathcal{M} \longrightarrow E_n \longrightarrow E_{n-1} \longrightarrow \dots \longrightarrow E_0 \longrightarrow s^*M \longrightarrow 0 \ .$$

PROPOSITION 1.- <u>Le S-module gradué $\mathcal{M}$ décrit ci-dessus possède les trois propriétés suivantes</u> :

a) <u>$\mathcal{M}$ est plat sur</u> A .

b) $X = (X_0,\dots,X_n)$ <u>est une suite -régulière.</u>

c) $[\mathcal{M}/\underline{X}\mathcal{M}]_j = ({}^j\wedge A^{n+1}) \otimes B_{n+1-j}$ <u>où</u> $B_k = \mathrm{Coker}(d_k)$.

La suite exacte $(*)$, regardée comme suite exacte de A-modules (i.e. on oublie le foncteur exact p_*.), montre que la dimension projective de $\mathcal{M}$ sur A est égale à zéro. D'où l'assertion a).

Comme $\mathcal{M}$ est un (n+1)-syzygie d'un S-module, toute suite S-régulière de longueur (n+1) est $\mathcal{M}$-régulière. Ce qui montre b).

Une présentation de $\mathcal{M}$ est fournie par la suite exacte

$$\bigoplus_{j=2}^{n+1} [p^*({}^j\wedge A^{n+1}) \otimes S(-j) \otimes L_{n+2-j}] \longrightarrow \bigoplus_{j=1}^{n+1} [p^*({}^j\wedge A^{n+1}) \otimes S(-j) \otimes L_{n+1-j}] \longrightarrow \mathcal{M} \longrightarrow 0 \ .$$

On a donc, pour tout j , une suite exacte :

$$({}^j\wedge A^{n+1}) \otimes L_{n+2-j} \longrightarrow ({}^j\wedge A^{n+1}) \otimes L_{n+1-j} \longrightarrow [\mathcal{M}/\underline{X}\mathcal{M}]_j \longrightarrow 0$$

et l'assertion c).

THEOREME 3.- <u>Soit</u> M <u>un module de dimension projective inférieure ou égale à un entier</u> n <u>sur un anneau noethérien</u> A . <u>Soient</u> $\mathcal{L}.$ <u>une résolution de longueur</u> n <u>de</u> M <u>et</u> $K.$ <u>le complexe de Koszul canonique sur l'anneau</u> $S = A[X_0,\dots,X_n]$. <u>Notons</u> p <u>le morphisme de projection</u> Spec S $\longrightarrow$ Spec A , $\mathcal{E}.$ <u>le complexe</u> $K. \otimes p^*\mathcal{L}.$ <u>et</u> $\mathcal{M}$ <u>le</u> <u>S-module gradué</u> $\mathrm{Ker}(E_n \longrightarrow E_{n-1})$.

<u>On a des isomorphismes fonctoriels, pour tout</u> $i \in \mathbb{Z}$,

$$[\mathrm{Tor}_i^S(S/\underline{X}, \mathcal{M} \otimes_A N)]_{n+1} = \mathrm{Tor}_i^A(M,N)$$

pour tout A-module N .

Pour $i = 0$, c'est le point c) de la proposition 1. Il reste à montrer que les foncteurs $[\mathrm{Tor}_i^S(S/\underline{X}, \mathfrak{M} \otimes_A .)]_{n+1}$ sont bien les foncteurs dérivés de $[S/\underline{X} \otimes (\mathfrak{M} \otimes_A .)]_{n+1}$, pour ceci, il suffit de vérifier :

a) $[\mathrm{Tor}_i^S(S/\underline{X}, \mathfrak{M} \otimes_A N]_{n+1} = 0$, si $i > 0$ et N est un A-module libre : c' le b) de la proposition 1 .

b) Les foncteurs $[\mathrm{Tor}_i^S(S/\underline{X}, \mathfrak{M} \otimes .)]_{n+1}$ donnent des suites exactes longues d'homologie pour une suite exacte courte de A-modules :

$$0 \longrightarrow N' \longrightarrow N \longrightarrow N'' \longrightarrow 0 \quad .$$

D'après a) de la proposition 1 , on a une suite exacte :

$$0 \longrightarrow \mathfrak{M} \otimes_A N' \longrightarrow \mathfrak{M} \otimes_A N \longrightarrow \mathfrak{M} \otimes_A N'' \longrightarrow 0 \quad .$$

On a donc le résultat voulu en prenant la partie homogène de degré $(n+1)$ de la suite exacte d'homologie de S-modules donnée par les foncteurs $\mathrm{Tor}_i^S(S/\underline{X},.)$.

Remarque.- On peut facilement voir que le théorème d'annulation des Tor_i n'est vrai en prenant seulement les parties homogènes de degré donné. Il serait intéres de pouvoir définir, pour une suite d'éléments $X_1,\ldots,X_n$ homogènes dans un annea gradué S , une catégorie de modules gradués qui satisfasse au théorème d'annula des $\mathrm{Tor}_i^S(S/\underline{X},.)$ pour une partie homogène de degré fixé.

- REFERENCES -

[1] M. AUSLANDER - Modules over unramified regular local rings, Proc. Int. Con of Maths 1962, 230-233.

[2] A. GROTHENDIECK - Eléments de géométrie algébrique, chapitre III, Publ. Ma I.H.E.S. vol. 17, 1963.

[3] D. MUMFORD - Lectures on curves on algebraic surfaces, Annals of Math. Stu n°59, Princeton Univ. Press.

[4] J.P. SERRE - Faisceaux algébriques cohérents, Annals of Math., 61, p. 197

521: G. Cherlin, Model Theoretic Algebra – Selected Topics. 34 pages. 1976.

522: C. O. Bloom and N. D. Kazarinoff, Short Wave Radiation lems in Inhomogeneous Media: Asymptotic Solutions. V. 104 s. 1976.

523: S. A. Albeverio and R. J. Høegh-Krohn, Mathematical ry of Feynman Path Integrals. IV, 139 pages. 1976.

524: Séminaire Pierre Lelong (Analyse) Année 1974/75. Edité . Lelong. V, 222 pages. 1976.

525: Structural Stability, the Theory of Catastrophes, and ications in the Sciences. Proceedings 1975. Edited by P. Hilton. 08 pages. 1976.

526: Probability in Banach Spaces. Proceedings 1975. Edited Beck. VI, 290 pages. 1976.

527: M. Denker, Ch. Grillenberger, and K. Sigmund, Ergodic ry on Compact Spaces. IV, 360 pages. 1976.

528: J. E. Humphreys, Ordinary and Modular Representations hevalley Groups. III, 127 pages. 1976.

529: J. Grandell, Doubly Stochastic Poisson Processes. X, pages. 1976.

530: S. S. Gelbart, Weil's Representation and the Spectrum e Metaplectic Group. VII, 140 pages. 1976.

531: Y.-C. Wong, The Topology of Uniform Convergence on r-Bounded Sets. VI, 163 pages. 1976.

532: Théorie Ergodique. Proceedings 1973/1974 Edité par Conze and M. S. Keane. VIII, 227 pages. 1976.

533: F. R. Cohen, T. J. Lada, and J. P. May, The Homology of ed Loop Spaces. IX, 490 pages. 1976.

534: C. Preston, Random Fields. V, 200 pages. 1976.

535: Singularités d'Applications Differentiables. Plans-sur-Bex. . Edité par O. Burlet et F. Ronga. V, 253 pages. 1976.

536: W. M. Schmidt, Equations over Finite Fields. An Elementary oach. IX, 267 pages. 1976.

537: Set Theory and Hierarchy Theory. Bierutowice, Poland . A Memorial Tribute to Andrzej Mostowski. Edited by W. Marek, ebrny and A. Zarach. XIII, 345 pages. 1976.

538: G. Fischer, Complex Analytic Geometry. VII, 201 pages.

539: A. Badrikian, J. F. C. Kingman et J. Kuelbs, Ecole d'Eté de abilités de Saint Flour V-1975. Edité par P.-L. Hennequin. IX, ages. 1976.

540: Categorical Topology, Proceedings 1975. Edited by E. Binz H. Herrlich. XV, 719 pages. 1976.

541: Measure Theory, Oberwolfach 1975. Proceedings. Edited Bellow and D. Kölzow. XIV, 430 pages. 1976.

542: D. A. Edwards and H. M. Hastings, Čech and Steenrod otopy Theories with Applications to Geometric Topology. VII, pages. 1976.

543: Nonlinear Operators and the Calculus of Variations, elles 1975. Edited by J. P. Gossez, E. J. Lami Dozo, J. Mawhin, . Waelbroeck, VII, 237 pages. 1976.

544: Robert P. Langlands, On the Functional Equations Satis- y Eisenstein Series. VII, 337 pages. 1976.

545: Noncommutative Ring Theory. Kent State 1975. Edited by Cozzens and F. L. Sandomierski. V, 212 pages. 1976.

546: K. Mahler, Lectures on Transcendental Numbers. Edited Completed by B. Diviš and W. J. Le Veque. XXI, 254 pages.

47: A. Mukherjea and N. A. Tserpes, Measures on Topological groups: Convolution Products and Random Walks. V, 197 s. 1976.

548: D. A. Hejhal, The Selberg Trace Formula for PSL (2, IR). ne I. VI, 516 pages. 1976.

549: Brauer Groups, Evanston 1975. Proceedings. Edited by linsky. V, 187 pages. 1976.

50: Proceedings of the Third Japan – USSR Symposium on ability Theory. Edited by G. Maruyama and J. V. Prokhorov. VI, ages. 1976.

Vol. 551: Algebraic K-Theory, Evanston 1976. Proceedings. Edited by M. R. Stein. XI, 409 pages. 1976.

Vol. 552: C. G. Gibson, K. Wirthmüller, A. A. du Plessis and E. J. N. Looijenga. Topological Stability of Smooth Mappings. V, 155 pages. 1976.

Vol. 553: M. Petrich, Categories of Algebraic Systems. Vector and Projective Spaces, Semigroups, Rings and Lattices. VIII, 217 pages. 1976.

Vol. 554: J. D. H. Smith, Mal'cev Varieties. VIII, 158 pages. 1976.

Vol. 555: M. Ishida, The Genus Fields of Algebraic Number Fields. VII, 116 pages. 1976.

Vol. 556: Approximation Theory. Bonn 1976. Proceedings. Edited by R. Schaback and K. Scherer. VII, 466 pages. 1976.

Vol. 557: W. Iberkleid and T. Petrie, Smooth S^1 Manifolds. III, 163 pages. 1976.

Vol. 558: B. Weisfeiler, On Construction and Identification of Graphs. XIV, 237 pages. 1976.

Vol. 559: J.-P. Caubet, Le Mouvement Brownien Relativiste. IX, 212 pages. 1976.

Vol. 560: Combinatorial Mathematics, IV, Proceedings 1975. Edited by L. R. A. Casse and W. D. Wallis. VII, 249 pages. 1976.

Vol. 561: Function Theoretic Methods for Partial Differential Equations. Darmstadt 1976. Proceedings. Edited by V. E. Meister, N. Weck and W. L. Wendland. XVIII, 520 pages. 1976.

Vol. 562: R. W. Goodman, Nilpotent Lie Groups: Structure and Applications to Analysis. X, 210 pages. 1976.

Vol. 563: Séminaire de Théorie du Potentiel. Paris, No. 2. Proceedings 1975–1976. Edited by F. Hirsch and G. Mokobodzki. VI, 292 pages. 1976.

Vol. 564: Ordinary and Partial Differential Equations, Dundee 1976. Proceedings. Edited by W. N. Everitt and B. D. Sleeman. XVIII, 551 pages. 1976.

Vol. 565: Turbulence and Navier Stokes Equations. Proceedings 1975. Edited by R. Temam. IX, 194 pages. 1976.

Vol. 566: Empirical Distributions and Processes. Oberwolfach 1976. Proceedings. Edited by P. Gaenssler and P. Révész. VII, 146 pages. 1976.

Vol. 567: Séminaire Bourbaki vol. 1975/76. Exposés 471–488. IV, 303 pages. 1977.

Vol. 568: R. E. Gaines and J. L. Mawhin, Coincidence Degree, and Nonlinear Differential Equations. V, 262 pages. 1977.

Vol. 569: Cohomologie Etale SGA 4½. Séminaire de Geometrie Algébrique du Bois-Marie. Edité par P. Deligne. V, 312 pages. 1977.

Vol. 570: Differential Geometrical Methods in Mathematical Physics, Bonn 1975. Proceedings. Edited by K. Bleuler and A. Reetz. VIII, 576 pages. 1977.

Vol. 571: Constructive Theory of Functions of Several Variables, Oberwolfach 1976. Proceedings. Edited by W. Schempp and K. Zeller. VI. 290 pages. 1977

Vol. 572: Sparse Matrix Techniques, Copenhagen 1976. Edited by V. A. Barker. V, 184 pages. 1977.

Vol. 573: Group Theory, Canberra 1975. Proceedings. Edited by R. A. Bryce, J. Cossey and M. F. Newman. VII, 146 pages. 1977.

Vol. 574: J. Moldestad, Computations in Higher Types. IV, 203 pages. 1977.

Vol. 575: K-Theory and Operator Algebras, Athens, Georgia 1975. Edited by B. B. Morrel and I. M. Singer. VI, 191 pages. 1977.

Vol. 576: V. S. Varadarajan, Harmonic Analysis on Real Reductive Groups. VI, 521 pages. 1977.

Vol. 577: J. P. May, E_∞ Ring Spaces and E_∞ Ring Spectra. IV, 268 pages. 1977.

Vol. 578: Séminaire Pierre Lelong (Analyse) Année 1975/76. Edité par P. Lelong. VI, 327 pages. 1977.

Vol. 579: Combinatoire et Représentation du Groupe Symétrique, Strasbourg 1976. Proceedings 1976. Edité par D. Foata. IV, 339 pages. 1977.

Vol. 580: C. Castaing and M. Valadier, Convex Analysis and Measurable Multifunctions. VIII, 278 pages. 1977.

Vol. 581: Séminaire de Probabilités XI, Université de Strasbourg. Proceedings 1975/1976. Edité par C. Dellacherie, P. A. Meyer et M. Weil. VI, 574 pages. 1977.

Vol. 582: J. M. G. Fell, Induced Representations and Banach *-Algebraic Bundles. IV, 349 pages. 1977.

Vol. 583: W. Hirsch, C. C. Pugh and M. Shub, Invariant Manifolds. IV, 149 pages. 1977.

Vol. 584: C. Brezinski, Accélération de la Convergence en Analyse Numérique. IV, 313 pages. 1977.

Vol. 585: T. A. Springer, Invariant Theory. VI, 112 pages. 1977.

Vol. 586: Séminaire d'Algèbre Paul Dubreil, Paris 1975–1976 (29ème Année). Edited by M. P. Malliavin. VI, 188 pages. 1977.

Vol. 587: Non-Commutative Harmonic Analysis. Proceedings 1976. Edited by J. Carmona and M. Vergne. IV, 240 pages. 1977.

Vol. 588: P. Molino, Théorie des G-Structures: Le Problème d'Equivalence. VI, 163 pages. 1977.

Vol. 589: Cohomologie l-adique et Fonctions L. Séminaire de Géométrie Algébrique du Bois-Marie 1965–66, SGA 5. Edité par L. Illusie. XII, 484 pages. 1977.

Vol. 590: H. Matsumoto, Analyse Harmonique dans les Systèmes de Tits Bornologiques de Type Affine. IV, 219 pages. 1977.

Vol. 591: G. A. Anderson, Surgery with Coefficients. VIII, 157 pages. 1977.

Vol. 592: D. Voigt, Induzierte Darstellungen in der Theorie der endlichen, algebraischen Gruppen. V, 413 Seiten. 1977.

Vol. 593: K. Barbey and H. König, Abstract Analytic Function Theory and Hardy Algebras. VIII, 260 pages. 1977.

Vol. 594: Singular Perturbations and Boundary Layer Theory, Lyon 1976. Edited by C. M. Brauner, B. Gay, and J. Mathieu. VIII, 539 pages. 1977.

Vol. 595: W. Hazod, Stetige Faltungshalbgruppen von Wahrscheinlichkeitsmaßen und erzeugende Distributionen. XIII, 157 Seiten. 1977.

Vol. 596: K. Deimling, Ordinary Differential Equations in Banach Spaces. VI, 137 pages. 1977.

Vol. 597: Geometry and Topology, Rio de Janeiro, July 1976. Proceedings. Edited by J. Palis and M. do Carmo. VI, 866 pages. 1977.

Vol. 598: J. Hoffmann-Jørgensen, T. M. Liggett et J. Neveu, Ecole d'Eté de Probabilités de Saint-Flour VI – 1976. Edité par P.-L. Hennequin. XII, 447 pages. 1977.

Vol. 599: Complex Analysis, Kentucky 1976. Proceedings. Edited by J. D. Buckholtz and T. J. Suffridge. X, 159 pages. 1977.

Vol. 600: W. Stoll, Value Distribution on Parabolic Spaces. VIII, 216 pages. 1977.

Vol. 601: Modular Functions of one Variable V, Bonn 1976. Proceedings. Edited by J.-P. Serre and D. B. Zagier. VI, 294 pages. 1977.

Vol. 602: J. P. Brezin, Harmonic Analysis on Compact Solvmanifolds. VIII, 179 pages. 1977.

Vol. 603: B. Moishezon, Complex Surfaces and Connected Sums of Complex Projective Planes. IV, 234 pages. 1977.

Vol. 604: Banach Spaces of Analytic Functions, Kent, Ohio 1976. Proceedings. Edited by J. Baker, C. Cleaver and Joseph Diestel. VI, 141 pages. 1977.

Vol. 605: Sario et al., Classification Theory of Riemannian Manifolds. XX, 498 pages. 1977.

Vol. 606: Mathematical Aspects of Finite Element Methods. Proceedings 1975. Edited by I. Galligani and E. Magenes. VI, 362 pages. 1977.

Vol. 607: M. Métivier, Reelle und Vektorwertige Quasimartingale und die Theorie der Stochastischen Integration. X, 310 Seiten. 1977.

Vol. 608: Bigard et al., Groupes et Anneaux Réticulés. XIV, 334 pages. 1977.

Vol. 609: General Topology and Its Relations to Modern Ana and Algebra IV. Proceedings 1976. Edited by J. Novák. XVIII, pages. 1977.

Vol. 610: G. Jensen, Higher Order Contact of Submanifold Homogeneous Spaces. XII, 154 pages. 1977.

Vol. 611: M. Makkai and G. E. Reyes, First Order Categorical L VIII, 301 pages. 1977.

Vol. 612: E. M. Kleinberg, Infinitary Combinatorics and the Axic Determinateness. VIII, 150 pages. 1977.

Vol. 613: E. Behrends et al., L^p-Structure in Real Banach Spa X, 108 pages. 1977.

Vol. 614: H. Yanagihara, Theory of Hopf Algebras Attached to G Schemes. VIII, 308 pages. 1977.

Vol. 615: Turbulence Seminar, Proceedings 1976/77. Edite P. Bernard and T. Ratiu. VI, 155 pages. 1977.

Vol. 616: Abelian Group Theory, 2nd New Mexico State Unive Conference, 1976. Proceedings. Edited by D. Arnold, R. Hunter E. Walker. X, 423 pages. 1977.

Vol. 617: K. J. Devlin, The Axiom of Constructibility: A Guide fo Mathematician. VIII, 96 pages. 1977.

Vol. 618: I. I. Hirschman, Jr. and D. E. Hughes, Extreme Eigen Va of Toeplitz Operators. VI, 145 pages. 1977.

Vol. 619: Set Theory and Hierarchy Theory V, Bierutowice 1 Edited by A. Lachlan, M. Srebrny, and A. Zarach. VIII, 358 pa 1977.

Vol. 620: H. Popp, Moduli Theory and Classification Theo Algebraic Varieties. VIII, 189 pages. 1977.

Vol. 621: Kauffman et al., The Deficiency Index Problem. VI, 112 pa 1977.

Vol. 622: Combinatorial Mathematics V, Melbourne 1976. Proc ings. Edited by C. Little. VIII, 213 pages. 1977.

Vol. 623: I. Erdelyi and R. Lange, Spectral Decomposition Banach Spaces. VIII, 122 pages. 1977.

Vol. 624: Y. Guivarc'h et al., Marches Aléatoires sur les Gro de Lie. VIII, 292 pages. 1977.

Vol. 625: J. P. Alexander et al., Odd Order Group Actions and Classification of Innerproducts. IV, 202 pages. 1977.

Vol. 626: Number Theory Day, New York 1976. Proceedings. E by M. B. Nathanson. VI, 241 pages. 1977.

Vol. 627: Modular Functions of One Variable VI, Bonn 1976. ceedings. Edited by J.-P. Serre and D. B. Zagier. VI, 339 pages.

Vol. 628: H. J. Baues, Obstruction Theory on the Homotopy C fication of Maps. XII, 387 pages. 1977.

Vol. 629: W.A. Coppel, Dichotomies in Stability Theory. VI, 98 pa 1978.

Vol. 630: Numerical Analysis, Proceedings, Biennial Confer Dundee 1977. Edited by G. A. Watson. XII, 199 pages. 1978.

Vol. 631: Numerical Treatment of Differential Equations. Procee 1976. Edited by R. Bulirsch, R. D. Grigorieff, and J. Schröd 219 pages. 1978.

Vol. 632: J.-F. Boutot, Schéma de Picard Local. X, 165 pages.

Vol. 633: N. R. Coleff and M. E. Herrera, Les Courants Rési Associés à une Forme Méromorphe. X, 211 pages. 1978.

Vol. 634: H. Kurke et al., Die Approximationseigenschaft lokaler R IV, 204 Seiten. 1978.

Vol. 635: T. Y. Lam, Serre's Conjecture. XVI, 227 pages.

Vol. 636: Journées de Statistique des Processus Stochastiques, noble 1977, Proceedings. Edité par Didier Dacunha-Castelle e nard Van Cutsem. VII, 202 pages. 1978.

Vol. 637: W. B. Jurkat, Meromorphe Differentialgleichunge 194 Seiten. 1978.

Vol. 638: P. Shanahan, The Atiyah-Singer Index Theorem, An duction. V, 224 pages. 1978.

Vol. 639: N. Adasch et al., Topological Vector Spaces. V, 125 pa 1978.